Hazardous Materials

Managing the Incident

FOURTH EDITION

Student Workbook

JoAnne Hildebrand

JONES & BARTLETT
LEARNING

World Headquarters
Jones & Bartlett Learning
5 Wall Street
Burlington, MA 01803
978-443-5000
info@jblearning.com
www.jblearning.com

Jones & Bartlett Learning books and products are available through most bookstores and online booksellers. To contact Jones & Bartlett Learning directly, call 800-832-0034, fax 978-443-8000, or visit our website, www.jblearning.com.

Substantial discounts on bulk quantities of Jones & Bartlett Learning publications are available to corporations, professional associations, and other qualified organizations. For details and specific discount information, contact the special sales department at Jones & Bartlett Learning via the above contact information or send an email to specialsales@jblearning.com.

Production Credits

Chief Executive Officer: Ty Field
President: James Homer
SVP, Editor-in-Chief: Michael Johnson
Executive Publisher: Kimberly Brophy
Vice President of Sales, Public Safety Group: Matthew Maniscalco
Director of Sales, Public Safety Group: Patricia Einstein
Senior Acquisitions Editor: Janet Maker
Editor: Alison Lozeau
Production Manager: Jenny L. Corriveau

Senior Marketing Manager: Brian Rooney
VP, Manufacturing and Inventory Control: Therese Connell
Composition: diacriTech
Cover Design: Kristin E. Parker
Director of Photo Research and Permissions: Amy Wrynn
Cover Image: © Courtesy of U.S. Coast Guard (top left) and Rob Schnepp (bottom left and right)
Printing and Binding: Courier Companies
Cover Printing: Courier Companies

To order this product, use ISBN: 978-1-4496-8829-5

Library of Congress Cataloging-in-Publication Data Not Available At Time of Printing

6048

Printed in the United States of America
17 16 15 14 13   10 9 8 7 6 5 4 3 2 1

Table of Contents

About the Author

JoAnne Fish Hildebrand is the former Academic Director of Fire Science at the University of Maryland University College (UMUC) with 22 years of service as a Collegiate Associate Professor. JoAnne chaired the USFA National Fire Academy (NFA) Degrees-at-a-Distance Higher Education Consortium. She is the author of the *Instructor's Guide to Hazardous Materials, Managing the Incident* (1st, 2nd, and 3rd editions) and the *Student Workbook to Hazardous Materials: Managing the Incident*, 2nd and 3rd editions. During her tenure at UMUC, she authored and taught a number of online courses and served on numerous Degrees-at-a-Distance Course Guide development teams at the National Fire Academy. She has taught, "Managerial Issues in Hazardous Materials," a 400-level course required for the major in fire science at UMUC, for more than 10 years using *Hazardous Materials: Managing the Incident* as the primary textbook.

JoAnne was a founding member of the NFPA Technical Committee on Occupational Safety and Health (NFPA 1500) and served on the committee for 9 years. She currently serves as an independent training and education consultant.

JoAnne received her M.A. and B.A., *Magna cum laude*, with the University of Maryland at College Park. She is a recipient of the University of Maryland University College's prestigious 2000 Stanley J. Drazek Teaching Excellence Award.

Introduction

The purpose of the Student Workbook is to provide a structured and enjoyable approach to studying the subject of hazardous materials incident response and management as presented in the textbook, *Hazardous Materials: Managing the Incident, Fourth Edition*, by Gregory G. Noll and Michael S. Hildebrand. The Student Workbook is particularly helpful to individuals who are preparing for a promotional examination based on the textbook, and for those who are attending a training class in order to qualify or requalify as a Hazardous Materials Technician or Hazardous Materials Incident Commander as described in the 2013 Edition of *NFPA 472, Standard for Competence of Responders to Hazardous Materials/Weapons of Mass Destruction Incidents.*

The Workbook is organized into 12 lessons, each corresponding to a chapter in the textbook and providing learning aids arranged in the following sections:

Chapter Orientation—This section is designed to familiarize you with the subjects being covered and to help you get a grasp of how the material is organized. The exercise helps you relate the material to your past experience and identify areas that will require more of your time or effort to study.

Learning Objectives—Learning objectives are statements that tell you what you are supposed to be able to do by the end of the lesson. This section helps you think about the chapter objectives in terms of their relevancy and importance in helping you reach your goals.

Abbreviations and Acronyms—Every occupation or field has a specialized vocabulary of abbreviations and acronyms, and the emergency response community is no exception. This section helps you learn the abbreviations and acronyms used in each chapter.

Study Session Overview—This section of the Workbook presents exercises to test your comprehension of the material in each chapter with multiple-choice, true/false, matching, fill-in-the-blank, and short-answer questions, similar to those you would find on a promotional examination, course quiz, or a standardized certification examination. Consult your instructor for access to The Student Workbook Answer Key.

Practice—This section of the lesson helps you achieve a deeper level of learning and retention that occurs when you have an opportunity to apply your knowledge or ask your own questions.

Important Terminology—This section of the Workbook helps you build your technical vocabulary and become a more effective communicator. Please note that not all chapters include a section on new terminology.

Study Group Activities—These activities are valuable learning tools for students who are able to meet in a study group, either face-to-face or by using online conferencing software. In addition to group activities, each chapter provides two real-world scenarios and a series of discussion questions sequentially arranged in an increasing level of complexity and abstraction. The first question is typically designed to help you apply basic "textbook" principles, and subsequent questions take the discussion beyond the basic technical issues and into the more difficult political, legal, and management related issues, often requiring you to make a personal judgment call on how to handle the problem. In most cases, there are no right or wrong answers, only defendable ones for the group to hammer out.

Summary and Review—This section provides questions that help you assess how well you have grasped the objectives of the chapter. If you have trouble answering any of the questions, you may need to review the material again.

Self-Evaluation—This section provides an opportunity to reflect on how well you feel you have mastered the material overall. You are directed to review all your work in the lesson and focus on areas where you see the need for improvement.

How to Use the Student Workbook

In order to use this Workbook most effectively, you will need a copy of the textbook, *Hazardous Materials: Managing the Incident, Fourth Edition*, by Gregory G. Noll and Michael S. Hildebrand. The textbook is divided into 12 chapters. Chapters 1 through 4 address preparing for the incident, with topics that include the development of a comprehensive system for managing the hazmat problem, health and safety issues and concerns, and the development and implementation of an emergency management organization. Chapters 5 through 12 pertain to implementing the hazmat response, beginning with an overview of the Eight Step Process© and with subsequent chapters covering each of the eight individual functions of the Eight Step Process©. The Eight Step Process© establishes a management structure that fits any size or level of hazmat response and provides an incident management framework with one specific goal—to maximize safety for emergency response personnel and the general public. The information contained within each of these chapters builds incrementally, in step-wise fashion.

Although it is optional, you should also have a copy of the *Field Operations Guide, Second Edition (FOG)* by Armando (Toby) S. Bevalaqua. The *FOG* includes detailed tactical checklists that follow the Eight Step Process©, a section on identification and recognition of containers, data cards on the top 50 hazardous materials and CBRNEs, as well as a matrix of WMD and drug lab precursor chemicals. The *FOG* is designed to be used at the incident scene and as a classroom reference guide to strategic and tactical decision-making.

Additional study aids can be found at http://www.fire.jbpub.com/hazmattech/

The Hazardous Materials Management System

Chapter Orientation

Open the textbook to the *Hazardous Materials Management System* chapter. When you have finished looking through the chapter, respond to the following items. Use the textbook as you jot down your comments in the spaces provided below.

1. In your own words, what makes a "systems approach" to hazardous materials management so important?

_____

_____

_____

_____

_____

2. Reflect on your current level of knowledge or background experience about the topics covered in this chapter. Where have you read about, learned about, or applied this knowledge in the past?

_____

_____

_____

_____

_____

3. Which sections or parts of this chapter strike you as looking especially interesting?

_____

_____

_____

_____

_____

4. Which particular subjects in this chapter are important for a person in your position to master?

_____

_____

_____

_____

_____

5. What do you predict will be the most difficult things for you to learn in this chapter?

_____

_____

_____

_____

_____

Learning Objectives

Examine the *Hazardous Materials Management System* chapter objectives and respond to the following question:

1. Which objectives in this chapter do you feel you can achieve right now, with a reasonable level of confidence?

_____

_____

_____

_____

_____

As you read the sections of the chapter that deal with the objectives you have identified above, make sure your ideas and knowledge base match those of the authors. If they do not match, you should examine how your current understanding of the material differs from that of the authors. Depending on the level at which you wish to master the subject, discrepancies will have to be rectified and gaps will need to be filled.

Abbreviations and Acronyms

The following abbreviations and acronyms are used in the *Hazardous Materials Management System* chapter:

ACP	Area Contingency Plan
ALS	Advanced Life Support
ANSI	American National Standards Institute
API	American Petroleum Institute
ASTM	American Society of Testing and Materials
CAA	Clean Air Act
CAER	Community Awareness and Emergency Response
CBRNE	Chemical, Biological, Radiological, Nuclear, and Explosive
CERCLA	Comprehensive Environmental Response, Compensation, and Liability Act
CFAT	Chemical Facility Anti-Terrorism
CFR	Code of Federal Regulations
CGA	Compressed Gas Association
CI/KR	Critical Infrastructure and Key Resources
CPG	*Comprehensive Preparedness Guide*
DHS	Department of Homeland Security
DOT	Department of Transportation
EHS	Extremely Hazardous Substance

EMS	Emergency Medical Services
EMT	Emergency Medical Technician
EOP	Emergency Operations Plan
EPA	Environmental Protection Agency
EPCRA	Emergency Planning and Community Right-to-Know Act
ERG	*Emergency Response Guidebook*
ERT	Emergency Response Team
FAA	Federal Aviation Administration
FBI	Federal Bureau of Investigation
FEMA	Federal Emergency Management Agency
FMECA	Failure Modes, Effects, and Criticality Analysis
FOG	Field Operations Guides
FOSC	Federal On-Scene Coordinator
FRA	Federal Railroad Administration
GHS	Globally Harmonized System
HAZCOM	Hazard Communication Regulations
HAZMAT	Hazardous Materials
HAZOP	Hazard and Operability Study
HAZWOPER	Hazardous Waste Operations and Emergency Response
HMRT	Hazardous Materials Response Team
HMT	Hazardous Materials Technician
IAEM	International Association of Emergency Managers
IC	Incident Commander
ICP	Integrated Contingency Plan
ICS	Incident Command System
IMS	Incident Management System
LEPC	Local Emergency Planning Committee
LGR	Local Government Reimbursement
NCP	National Contingency Plan
NFPA	National Fire Protection Association
NIJ	National Institute of Justice
NRC	National Response Center
NRT	National Response Team
NTSB	National Transportation Safety Board
OHME	Office of Hazardous Materials Enforcement
OHMS	Office of Hazardous Materials Safety
OPA	Oil Pollution Act of 1990
OPS	Office of Pipeline Safety
OSC	On-Scene Coordinator
OSHA	Occupational Safety and Health Administration
PHMSA	Pipeline and Hazardous Materials Safety Administration
PIO	Public Information Officer
PSM	Process Safety Management

RCRA	Resource Conservation and Recovery Act
RMP	Risk Management Program
RP	Responsible Party
RPM	Remedial Project Manager
RRT	Regional Response Team
SARA	Superfund Amendments and Reauthorization Act of 1986
SDS	Safety Data Sheet
SEI	Safety Equipment Institute
SERC	State Emergency Response Commission
SOP/SOG	Standard Operating Procedures/Guidelines
TRANSCAER	Transportation Community Awareness and Emergency Response
TSA	Transportation Security Administration
USCG	United States Coast Guard
WMD	Weapons of Mass Destruction

Abbreviations and Acronyms Exercise

For each of the following sentences, write in the correct abbreviation or acronym (from the preceding list) so that that sentence makes sense. Use each abbreviation or acronym only once.

1. The _____ is a massive publication containing all the rules and regulations enforced by the various federal departments and agencies.

2. Also known as the Emergency Planning and Community Right-to-Know Act, SARA, Title III requires the establishment of _____ and LEPCs.

3. The highly successful Wally Wise Guy™ public education program was developed by the Deer Park, TX, _____.

4. PHMSA's _____ is revised and published on a three-year cycle.

5. The term "MSDS" is changing to the term _____ under the Globally Harmonized System of Classification and Labeling of Chemicals.

6. The EPA's _____ program can provide up to $25,000 per incident to local governments that do not have funds available to pay for response actions.

7. The job of Hazardous Materials Specialists is to respond with and provide support to the _____.

8. The risk management program regulation is similar in scope to the OSHA _____ standard, with the primary focus being community safety as compared to employee safety.

9. A weapon involving toxic or poisonous chemicals is an example of a(n) _____.

10. In the aftermath of the terrorist attacks of September 11, 2001, numerous regulations and programs pertaining to the security and protection of hazardous materials _____ have been enacted.

Crossword: Organization Abbreviations

Use the clues below to solve the crossword.

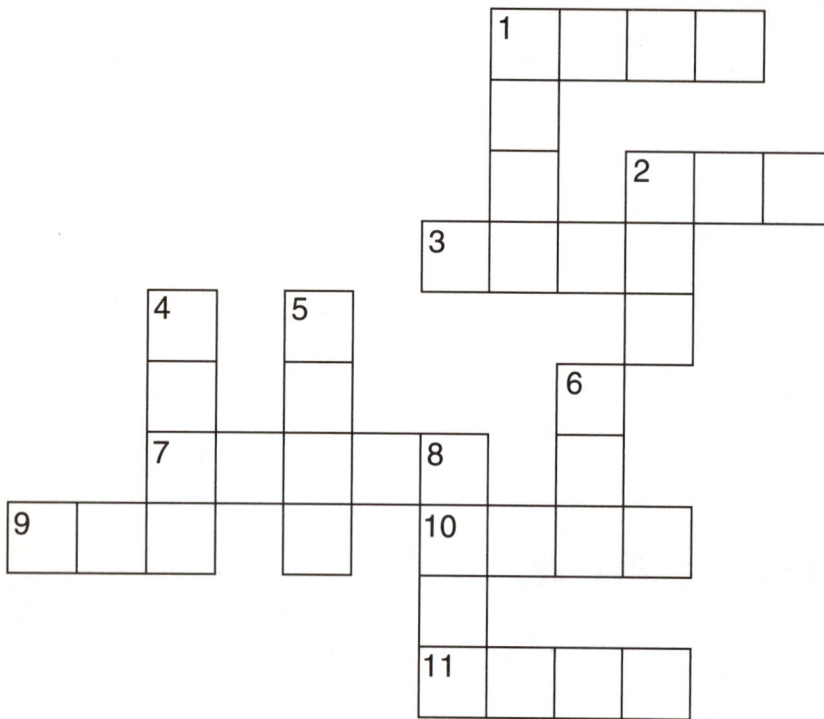

Across

1. This agency is responsible for the management and coordination of PHMSA's hazardous materials inspection and enforcement program.

2. This agency assumes the role of federal OSC during the emergency response phase if the incident is a terrorism-related event.

3. HAZWOPER was written and is enforced by this agency.

7. This agency administers DOT's national regulatory program to ensure the safe transportation of natural gas, petroleum, and other hazardous materials by pipeline.

9. This organization develops and promotes safety standards and safe practices in the industrial gas industry.

10. This independent federal agency conducts hazardous materials studies and accident investigation reports.

11. This nonprofit organization represents emergency managers.

Down

1. This agency is responsible for all hazardous materials transportation regulations except bulk shipment by ship or barge.

2. This agency enforces regulations relating to hazardous materials shipments on domestic and foreign carriers operating at U.S. airports.

4. This organization is recognized for its role in developing recommended practices that affect fire safety and hazmat operations.

5. This agency offers a web-based, free-of-charge independent study course entitled *IS-55: Household Hazardous Materials*.

6. This agency established the Chemical Facility Anti-Terrorism (CFAT) standards.

8. This is one of several important standards-writing bodies that affect the hazardous materials emergency response community.

Study Session Overview

Use the textbook to help you answer the questions or work from memory.

1. How does Benner's definition of a hazardous material differ from the DOT definition?

   _____

   _____

   _____

   _____

   _____

2. In your own words, explain the difference between hazmat laws and hazmat regulations. Give an example of each.

   _____

   _____

   _____

   _____

   _____

3. Give an example of a hazmat-related voluntary consensus standard and explain briefly how it impacts your emergency service organization.

   _____

   _____

   _____

   _____

   _____

4. How could failure to comply with a standard of care create a legal problem for an emergency services organization?

   _____

   _____

   _____

   _____

5. Briefly describe a plausible scenario that would be considered a Level II Hazardous Materials Incident in your community.

   _____

   _____

   _____

   _____

6. In general terms, which safety issues are associated with clean-up operations?

_____

_____

_____

_____

_____

_____

Self-Test

Use the textbook to help you answer the questions or work from memory.

1. The following statements are all true *except*:
   a. Hazardous materials can be found virtually anywhere.
   b. Hazardous materials can be used as weapons for criminal or terrorist purposes.
   c. All government agencies use the same standard definition of hazardous materials.
   d. The DOT is one of several federal agencies that regulate hazardous materials.

2. In which context is the term "dangerous goods" most likely to be used?
   a. Workplace exposures
   b. Agricultural chemicals
   c. Interstate transportation
   d. International transportation

3. Which of the following regulatory system components is enacted through legislative action?
   a. Laws
   b. Regulations
   c. Rules
   d. Voluntary consensus standards

4. When a federal agency or state or municipal government adopts a consensus standard by reference, the document becomes a:
   a. recommended practice.
   b. performance-based code.
   c. regulation.
   d. formal interpretation.

5. Which of the following laws is known as the "Superfund" law?
   a. Oil Pollution Act of 1990
   b. Clean Air Act
   c. Resource Conservation and Recovery Act (1976)
   d. Comprehensive Environmental Response Compensation and Liability Act (1980)

6. Hazardous Waste Operations and Emergency Response (29 CFR 1910.120), also known as HAZWOPER, is enforced in many states by:

    a. FEMA.

    b. OSHA.

    c. EPA.

    d. DOT.

7. HAZWOPER establishes requirements for all of the following *except*:

    a. medical surveillance programs.

    b. emergency response procedures.

    c. single-point-of-contact reporting.

    d. post-emergency termination procedures.

8. As identified in OSHA 1910.120(q)(6), the following responders are all trained to be capable of implementing the local Emergency Operations Plan *except*:

    a. Specialist Employees.

    b. Hazardous Materials Technician.

    c. Hazardous Materials Specialists.

    d. On-Scene Incident Commander.

9. Which of the following groups is responsible for developing and maintaining the state's emergency response plan?

    a. Regional Response Team

    b. Chemical Emergency Preparedness and Prevention Office

    c. Local Emergency Planning Committee

    d. State Emergency Response Commission

10. Which of the following groups is responsible for coordinating the Community Right-to-Know aspects of SARA, Title III?

    a. Local Emergency Planning Committee

    b. State Department of the Environment

    c. Office of Hazardous Materials Safety

    d. Chemical Emergency Preparedness and Prevention Office

11. Which of the following organizations assumes the role of Federal On-Scene Coordinator (OSC) when the incident is a terrorism-related event?

    a. U.S. Coast Guard

    b. National Response Team

    c. Federal Bureau of Investigation

    d. Defense Threat Reduction Agency

12. Which of the following organizations develops voluntary consensus standards concerning hazardous materials?

    a. National Fire Protection Agency (NFPA)

    b. Compressed Gas Association (CGA)

    c. American Petroleum Institute (API)

    d. All the above

13. There are four key elements in a hazardous materials management systems approach: 1) planning and preparedness; 2) prevention; 3) response; and 4):

    a. inspection and enforcement.

    b. clean-up and recovery.

    c. critique.

    d. event tree analysis.

14. Risk analysis assesses two considerations: 1) the probability or likelihood of an accidental release, and 2):

    a. the adequacy of response capabilities.

    b. the size/extent of vulnerability zones.

    c. the actual consequences that might occur.

    d. the potential environmental impact.

15. Both OSHA 1910.120 and NFPA 472 recommend that HMRT personnel be trained to:

    a. On-Scene Incident Commander level.

    b. Hazardous Materials Specialist level.

    c. Hazardous Materials Technician level.

    d. First Responder Operations level.

16. True / False: Federal installations and military bases are exempt from EPA's right-to-know regulations.

17. True / False: Each of the 50 states and the U.S. territories maintains an enforcement agency that has responsibility for hazardous materials.

18. True / False: The purpose of NFPA 472 is to specify minimum competencies for those who will respond to hazardous materials incidents.

19. True / False: "Standard of Care" represents the minimum accepted level of hazardous materials emergency service that should be provided regardless of location or situation.

20. True / False: Almost all hazardous materials facilities, containers, and processes are designed and constructed to some standard.

Practice

1. Visit your LEPC web site. Does it inform citizens in your county how they can find which chemical risks are present in their community?

   _____

   _____

   _____

   _____

2. Conduct an informal survey of your co-workers about the "standard of care" and what it means to them and the organization. Do your findings identify any misunderstandings about the concept in relation to the discussion provided in this chapter?

   _____

   _____

   _____

   _____

3. Read *NFPA 472, Standard for Competence of Responders to Hazardous Materials/Weapons of Mass Destruction Incidents* (2013 edition). This document can be ordered from the National Fire Protection Association or can be accessed for online review (see www.nfpa.org). Compare and contrast the duties and responsibilities of Hazardous Materials Technicians with those of Incident Commanders as described in the standard.

_____

_____

_____

_____

4. Take the Emergency Management Institute's interactive web-based course, entitled IS-340 *Hazardous Materials Prevention*, located at http://training.fema.gov/EMIWeb/IS/is340.asp or by searching the EMI web site. Test your knowledge by taking the final examination. Jot down at least three things you learned in this course about hazmat emergency planning.

_____

_____

_____

_____

Important Terminology

The following are all important terms that you should know.

Hazardous materials
Hazardous substances
Extremely hazardous substances (EHSs)
Hazardous chemicals
Hazardous wastes
Dangerous goods

On the line next to each term in Column A, print the letter of its definition from Column B.

Column A	Column B
Terminology	**Definitions**
_____ **1.** Hazardous materials	**A.** Chemicals determined by the EPA to be extremely hazardous to a community during an emergency spill or release as a result of their toxicities and physical/chemical properties.
_____ **2.** Hazardous substances	**B.** In international transportation, the term used to refer to hazardous materials.
_____ **3.** Extremely hazardous substances (EHSs)	**C.** Any substance designated under the Clean Water Act and the Comprehensive Environmental Response, Compensation and Liability Act (CERCLA) as posing a threat to waterways and the environment when released.
_____ **4.** Hazardous chemicals	**D.** Any chemical that would be a risk to employees if exposed in the workplace.
_____ **5.** Hazardous wastes	**E.** Any substance or material in any form or quantity that poses an unreasonable risk to safety and health and property when transported in commerce.
_____ **6.** Dangerous goods	**F.** Discarded materials regulated by the EPA because of public health and safety concerns. Regulatory authority is granted under the Resource Conservation and Recovery Act (RCRA).

Study Group Activity

This activity is a "scavenger hunt"–type exercise that helps you learn more about the NFPA and its voluntary consensus standards process.

Access the NFPA web site at www.nfpa.org to answer the following questions.

1. What is the name of the NFPA's main blog?

2. Who is eligible to submit a proposal to the NFPA for a new safety-related project?

3. How many individuals are appointed to the Standards Council, and who appoints them?

4. Which NFPA Committee is responsible for NFPA 472, *Professional Competence of Responders to Hazardous Materials Incidents?*

5. How many Committee member classifications are there?

6. Which Committee members may have an alternate member?

7. All NFPA consensus codes and standards are developed under the procedures of another well-known standards organization. Which one?

8. Several major conferences are hosted by NFPA annually. At which is a technical committee report session held that offers members an opportunity to vote on NFPA code and standard actions and offers the public the opportunity to voice their opinions on those actions? During which month is it held?

9. The NFPA has a "Student Member" membership category. What are the criteria for becoming a student member? Can student members vote?

10. The NFPA provides an educational grant for hazardous materials response teams. What is the name of the grant, in what amount is each grant, who can apply, and what must the grant be used for?

11. What are the key changes to the 2013 edition of NFPA 472, *Standard for Competence of Responders to Hazardous Materials/WMD Incidents*, from the 2008 edition? What has been added?

12. In which chapter of the 2013 edition of NFPA 472 are the competencies of Hazardous Materials Technicians covered? In which chapter are the competencies of the Incident Commander covered?

Study Group Learning Through Inquiry Scenario 1–1

You are the emergency response coordinator for an agricultural chemical manufacturing facility. Your company recently hired a new safety manager who is a Certified Safety Professional (CSP), but does not have any background in emergency response.

During the initial meeting with your new boss, she explains that she recently attended a national conference that conducted a simulated court case involving an emergency response organization that received an OSHA citation for its failure to have proper hazmat training and adequate site safety procedures. Based on what she learned at the conference, your boss is concerned that the plant is not in total compliance with the OSHA 29 CFR 1910.120 requirements. (Refer to background information on CFR 1910.120 in the *Hazardous Materials Management System* chapter.)

Your supervisor requests a briefing on the plant's hazmat response and training program. As part of the briefing, you outline the following points:

- The Emergency Response Team (ERT) is an integrated unit that responds to fires, hazmat spills, and basic medical emergencies.
- All of your ERT members are currently trained to the HAZWOPER First Responder Operations level.
- There is no formal Hazardous Materials Response Team (HMRT) on the complex because you rely on the local fire department, which does have an HMRT. (If you are unfamiliar with the concept of an HMRT, review the material in this chapter.)
- ERT members are trained to the Advanced Exterior Firefighting level, as identified in NFPA 600, *Industrial Fire Brigades*. The plant manufactures pesticides and the most significant fire problems involve flammable and combustible liquids as well as flammable gases.
- Your command officers have not received any training beyond the First Responder at the Operations level and an 8-hour Incident Command Course provided by an outside contractor.
- The ERT does not have a formal safety officer.

Using the background information provided and the information discussed in Chapter 1 answer the following questions:

1. Based on your understanding of 29 CFR 1910.120 (HAZWOPER), is your ERT in compliance? If you believe it is in compliance, support your position using the material provided in the *Hazardous Materials Management System*.

2. If you are uncertain whether your ERT is in compliance with 29 CFR 1910.120, how would you determine that it is? Explain the basic plan that you would develop for determining that you are in compliance.

3. If your chemical plant were inspected by OSHA, do you think the company would be issued a citation? If you believe citations would be issued, do you think citations would be issued based on your current level of ERT training and level of OSHA compliance under 29 CFR 1910.120? If so, why?

4. Assuming that some managers were not in compliance with the Incident Commander training requirements, how would you handle a situation where your new supervisor vaguely suggests that you could modify certain training records so that they were in compliance—in other words, knowingly falsify the training records. If you answered that you would refuse to falsify the training records, would you do so if you were verbally ordered to make such changes? How would you react under the same circumstances if you had 1 year to go before retirement? What if you were up for a promotion 1 year before retirement and the increase in salary would raise your pension?

Study Group Learning Through Inquiry Scenario 1–2

You were the incident commander at a multiple-alarm warehouse fire that occurred 3 years ago in your community. The warehouse stored a wide range of petroleum-based paints and paint thinners. You made the decision to attack and extinguish the fire; however, the plan failed due to inadequate water supply. Unfortunately, there was significant runoff and pollution associated with this incident, and several nearby drinking water wells were contaminated as a result of the runoff.

The affected residents have joined in a class-action lawsuit against your organization. One of the elements of the complaint is that you acted irresponsibly by attempting to extinguish the fire. The plaintiff's attorney has taken a deposition from an expert who has stated that a fire of this type should have been handled as a controlled burn. In other words, the expert is saying that you made the wrong decision.

Your organization's attorney informs you that an important element in the case will involve a concept known as the "standard of care." He warns you that you will have to give a deposition. You will probably be asked many difficult questions concerning whether you and your organization met the minimum level of accepted hazmat service expected for an emergency response organization of similar size. The attorney representing the residents will attempt to show that (1) you made the wrong decision and (2) that you and your organization violated a standard of care. Review the term "standard of care" in the *Hazardous Materials Management System* chapter before answering the following questions.

1. As an incident commander, how do you personally define the standard of care for an emergency response organization? List the primary standards, laws, or regulations that you feel you will be measured against to establish the minimum standard of care.

2. Using your own definition of a standard of care, do you feel that a "career," or fully paid, emergency response organization should be held to a higher standard of care than an all-volunteer organization? For example, does the standard of care differ for a career fire department versus a volunteer fire department? If you answered "yes," why should a volunteer be held to more or less of a standard than a career individual or organization? Which standard determines whether an individual is a "professional"?

3. Should an organized Hazardous Materials Response Team as defined by 29 CFR 1910.120 be held to a higher standard of care than a multipurpose emergency response organization (e.g., fire department engine company that meets NFPA 1001 or an industrial fire brigade that meets NFPA 600)?

Summary and Review

1. Identify the primary target audience for this text.

2. True / False: The scope of the textbook is such that readers must have advanced-level training in chemistry.

3. EHS is an abbreviation for:

4. What is the U.S. Department of Transportation's (DOT) definition of hazardous materials?

5. Voluntary consensus standards are normally developed through the actions of:
   a. municipal government.
   b. trade associations.
   c. local emergency planning committees.
   d. Congress.

6. CERCLA—the Comprehensive Environmental Response, Compensation and Liability Act (1980), also known as Superfund—requires those individuals responsible for the release of a hazardous material (commonly referred to as the responsible party) above a specified "reportable quantity" to notify the:
   a. National Response Center (NRC).
   b. State Emergency Response Commission.
   c. Occupational Safety and Health Administration (OSHA).
   d. Office of Hazardous Materials Enforcement.

7. SARA Title III is also known as the:
   a. Oil Pollution Act.
   b. Clean Air Act.
   c. Federal Water Pollution Control Act.
   d. Emergency Planning and Community Right-to-Know Act (EPCRA).

8. Hazardous Waste Operations and Emergency Response (29 CFR 1910.120), also known as HAZWOPER, establishes important requirements for both industry and public safety organizations that respond to hazmat or hazardous waste emergencies. Requirements cover the following areas:
   • Hazmat emergency response plan
   • Emergency response procedures, including the establishment of an Incident Management System (IMS), the use of a buddy system with backup personnel, and the establishment of the role of safety officer
   • Specific training requirements covering instructors and both initial and refresher training

   There are two other requirements. Identify one:

9. According to the specific levels of competency and associated training requirements identified within OSHA 1910.120(q)(6), First Responder–Operations personnel shall have sufficient training or experience to demonstrate objectively all of the following competencies *except*:
   a. capable of implementing the local emergency operations plan.
   b. an understanding of basic hazardous materials terms.
   c. knowledge of how to implement basic decontamination measures.
   d. knowledge of basic hazard and risk assessment techniques.

10. Among the membership of Local Emergency Planning Committees (LEPC) are representatives from the following groups:
   • Elected state and local officials
   • Fire department
   • Law enforcement

   Identify three other groups:

11. The Local Emergency Planning Committee (LEPC) is responsible for all of the following *except*:
   a. develop, regularly test, and exercise the hazmat emergency operations plan.

   b. manage the state's hazardous materials inspection and enforcement program within the community.

   c. receive and manage hazmat facility reporting information.

   d. conduct a hazards analysis of hazmat facilities and transportation corridors.

12. When the National Response Team or Regional Response Team is activated for a federal response to an oil spill, hazmat, or terrorism event, a Federal On-Scene Coordinator (FOSC) will be designated to coordinate the overall response. For hazmat incidents, the FOSC will represent either EPA or the _____ based on the location of the incident.
   a. Department of Transportation (DOT)

   b. U.S. Coast Guard (USCG)

   c. Federal Bureau of Investigation (FBI)

   d. National Transportation Safety Board (NTSB)

13. The purpose of _____ is to specify minimum requirements of competence and to enhance the safety and protection of response personnel and all components of the EMS system.
   a. NFPA 471

   b. NFPA 473

   c. NFPA 1991

   d. NFPA 1994

14. NFPA 472 defines three levels of:
   a. Hazardous Materials Technician.

   b. Hazardous Materials Safety Officer.

   c. Specialist employee.

   d. Hazmat Group Supervisor.

15. The standard of care is established on two bases:
   • Existing laws and regulations
   • Voluntary consensus standards and recommended practices

   The standard of care is *also* influenced by:

16. Identify the four key elements in a hazardous materials management systems approach:

17. There are four components of a hazards analysis program:
   1. Hazards identification

   2. Vulnerability analysis

   3. _____

   4. Emergency response resources evolution

18. Clean-up and recovery operations are designed to (1) clean up/remove the hazmat spill or release, and (2):

Self-Evaluation

Review all of your work in this lesson and note your stronger and weaker areas.

Overall I feel I did (very well / well / fair / not so well) on the acronyms and abbreviations exercise and crossword puzzle.

Overall I feel I did (very well / well / fair / not so well) on the self-test questions.

Overall I feel did (very well / well / fair / not so well) on the terminology (matching) exercise.

Overall I feel I did (very well / well / fair / not so well) on the summary and review questions.

List two areas in this chapter in which you feel you could improve your skill or knowledge level:

1. _____
_____
_____
_____

2. _____
_____
_____
_____

Consider the following self-evaluation questions as they pertain to this chapter:

Am I taking effective notes?

Am I dedicating enough quality time to my studies?

Is anything distracting my focus?

Was any part of this chapter too advanced for me?

Did I find that I don't have enough background experience to sufficiently grasp certain subject areas?

For areas in which I did particularly well, was it because I'm particularly interested in that subject matter? How so?

Were some things easier to learn because I have prior experience in learning or working with the concepts or principles?

Did I find that certain portions of the textbook seem to be better organized and effective in explaining key points?

Health and Safety

Chapter Orientation

Open the textbook to the *Health and Safety* chapter. When you have finished looking through the chapter, respond to the following items. Use the textbook as you jot down your comments in the spaces provided below.

1. In your own words, what makes learning about the relationship of toxicity and exposure so important?

_____

_____

_____

_____

_____

2. Reflect on your current level of knowledge or background experience about the topics covered in this chapter. Where have you read about, learned about, or applied this knowledge in the past?

_____

_____

_____

_____

_____

3. Which sections or parts of this chapter strike you as looking especially interesting?

_____

_____

_____

_____

_____

4. Which particular subjects in this chapter are important for a person in your position to master?

_____

_____

_____

_____

5. What do you predict will be the most difficult things for you to learn in this chapter?

_____

_____

_____

_____

Learning Objectives

Examine the *Health and Safety* chapter objectives and respond to the following question:

1. Which objectives in this chapter do you feel you can achieve right now, with a reasonable level of confidence?

_____

_____

_____

_____

_____

As you read the sections of the chapter that deal with the objectives you have identified above, make sure your ideas and knowledge base match those of the authors. If they do not match, you should examine how your current understanding of the material differs from that of the authors. Depending on the level at which you wish to master the subject, discrepancies will have to be rectified and gaps will need to be filled.

Abbreviations and Acronyms

The following abbreviations and acronyms are used in the *Health and Safety* chapter:

ACGIH	American Conference of Governmental Industrial Hygienists
AEGL	*Acute Emergency Exposure Guidelines*
AIHA	American Industrial Hygiene Association
ALARA	As Low as Reasonably Achievable
BEI	Biological Exposure Indices
BP	Blood Pressure
Bq	Becquerel
CGy	Centi-Gray
Ci	Curie
CNS	Central Nervous System
CPC	Chemical Protective Clothing
CPR	Cardiopulmonary Resuscitation
dBA	Decibels on the A-Weighted Scale
DoD	Department of Defense
dps	Disintegrations per Second
ECBC	Edgewood Chemical Biological Center

ECC	Emergency Cardiac Care
EKG	Electrocardiogram
ERPG	*Emergency Response Planning Guideline*
Gy	Gray
HCN	Hydrogen Cyanide
IARC	International Agency for Research on Cancer
IDLH	Immediately Dangerous to Life or Health
kg	Kilogram
lb	Pounds
LC	Lethal Concentration
LD	Lethal Dose
LOC	Level of Concern
m^3	Cubic Meter
µg	Microgram
mg	Milligram
mm Hg	Millimeters of Mercury
MSv	Milli-Sievert
NIOSH	National Institute of Occupational Safety and Health
NTP	National Toxicology Program
PAR	Personal Accountability Review
PCBs	Polychlorinated Biphenyls
PCM	Phase-Liquid Change Material
PEL	Permissible Exposure Limit
ppb	Parts per Billion
PPE	Personal Protective Equipment
ppm	Parts per Million
R	Roentgen
RAD	Radiation Absorbed Dose
Rehab	Rehabilitation Area
REL	Recommended Exposure Levels
rem	Roentgen Equivalent for Man
SBCCOM	U.S. Army Soldier and Biological Chemical Command
SI	International Systems Units
SOP	Standard Operating Procedures
STEL	Short-Term Exposure Limit
Sv	Sievert
TLV	Threshold Limit Value
TLV/C	Threshold Limit Value/Ceiling
TLV/STEL	Threshold Limit Value/Short-Term Exposure Limit
TLV/TWA	Threshold Limit Value/Time-Weighted Average

Abbreviations and Acronyms Exercise

For each of the following sentences, write in the correct abbreviation or acronym (from the preceding list) so that that sentence makes sense. Use each abbreviation or acronym only once.

1. There are some chemicals that cannot be excreted from the body, such as hydrogen fluoride, which accumulates in the bones, and _____, which accumulate in body fat.

2. Two units of measurement are commonly used for determining the relative toxicity of a chemical substance or compound: lethal dose (LD) and _____.

3. Experience over the last 30 years has shown that responders wearing _____ are more likely to be injured as a result of heat stress than a chemical exposure.

4. Worn under PPE, _____ operates under the principle of conductive heat cooling.

5. Strengths of radioactive sources are measured using the SI unit Bq; however, the old system of _____ may still be used in medical and technical practices.

6. The acronym _____ is based on the assumption that any exposure to ionizing radiation carries with it some risk.

7. Referred to as "eagles," _____ are intended to provide uniform exposure guidelines for the general public for a single short-term exposure.

8. OSHA regulations require that hearing protection be provided whenever noise levels exceed 85 _____.

9. One of the principal tasks of the _____ is to recommend TLVs for workplace exposure to chemicals.

10. Developed by the AIHA as an emergency planning tool for public protective action options, _____ have been developed for approximately 100-plus chemicals.

11. The abbreviation for electrocardiogram is _____.

12. The abbreviation for Sievert is _____.

Study Session Overview

Use the textbook to help you answer the questions or work from memory.

1. Using the template below, substitute another common substance for bourbon (which was used in the chapter) and indicate the acute and chronic effects as related to dose (examples include sunlight, cigarette smoke, coffee).

Dose	Acute Effect	Chronic Effect

2. In your own words, explain the difference between a "local effect" and a "systemic effect" of a hazmat exposure. Give an example of each.

_____

_____

_____

_____

_____

3. How can you estimate the IDLH of a toxic chemical if you know its TLV/TWA?

_____

_____

_____

_____

_____

4. In your own words, explain time, distance, and shielding as they relate to exposure to radioactive materials.

_____

_____

_____

_____

_____

5. Which conditions promote or have an effect on heat stress?

_____

_____

_____

_____

_____

6. Under which conditions can the ABCs of emergency treatment be administered to a contaminated victim (i.e., a victim who has not been decontaminated).

_____

_____

_____

_____

7. Describe an ideal area in which to situate a rehabilitation area.

_____

_____

_____

_____

Self-Test

Use the textbook to help you answer the questions or work from memory.

1. Exposure + _____ = Health Hazard

   a. Concentration

   b. Dose

   c. Toxicity

   d. Chemical agent

2. Absorption through the _____ is one of the fastest means of exposure.

   a. eyes

   b. palms and fingertips

   c. toxicity

   d. chemical agent

3. What is the target organ of hepatotoxins?

   a. Kidneys

   b. Liver

   c. Bones

   d. Blood system

4. Lead and organophosphate pesticides are examples of:

   a. nephrotoxins.

   b. hematotoxins.

   c. teratogens.

   d. neurotoxins.

5. Dose = Concentration × _____

   a. Time

   b. Toxicity

   c. Effect

   d. Rate of Absorption

6. Which of the following is a measurement commonly cited for determining the relative toxicity of a chemical?

   a. pH

   b. PEL

   c. LD_{50}

   d. ppm

7. Which of the following is a physical indicator of likely IDLH conditions?

   a. Visible vapor cloud

   b. Confined spaces

   c. Dead birds or discolored foliage

   d. All of the above

8. Which of the following is an example of ionizing radiation?

   a. Microwaves

   b. Lasers

   c. X-rays

   d. Infrared waves

9. In the United States, radioactive materials dose limits for workers performing emergency services are provided in units of:

   a. Ci.

   b. rem.

   c. Gy.

   d. Sv.

10. The following are all symptoms of heat stroke *except*:

   a. dry, hot skin.

   b. muscle cramps.

   c. weakness.

   d. full rapid pulse.

11. Wet clothing extracts heat from the body up to _____ faster than dry clothing.

   a. 5 times

   b. 20 times

   c. 50 times

   d. 240 times

12. Which of the following is an effect of excessive noise levels?

   a. Personnel being annoyed, startled, or distracted

   b. Physical damage to ears, pain, and temporary and/or permanent hearing loss

   c. Interference with communications, which may limit the ability of ERP to warn of danger or enforce proper safety precautions

   d. All of the above

13. There are five components of a medical surveillance program for hazmat responders, including pre-employment screening, periodic medical examination, emergency treatment, recordkeeping and review, and:

   a. nonemergency treatment.

   b. physician referral.

   c. physical fitness program.

   d. wellness program.

14. Which of the following is an advantage of using operational checklists to meet the site safety requirements?

   a. Ability to ensure that specific organizational guidelines and SOPs are followed

   b. Ability to track activities and performance

   c. Ability to document the plan of action and decision-making process

   d. All of the above

15. When is post-entry medical monitoring performed?

   a. Before leaving the hot zone

   b. Following decontamination

   c. During debriefing

   d. Within 24 hours of entry

16. Where should the rehabilitation area be located?

   a. Near the incident command post to facilitate communications

   b. In the staging area where it is easily accessible by EMS units

   c. In the warm zone to facilitate prompt reentry following rehab

   d. In a location that provides physical rest by allowing the body to recuperate from the hazards and demands of the emergency

17. True / False: Skin absorption can occur with no sensation to the skin itself.

18. True / False: The lower the reported concentration, the more toxic the material.

19. True / False: Personal protective clothing provides moderate protection against gamma radiation.

20. True / False: Both the incident safety officer and all assistant safety officers must have authority to stop any operations that are deemed unsafe.

Practice

1. The human body can be subject to seven types of harm events, listed below and discussed in this chapter. Using other sources, locate actual events in which an emergency responder was harmed by each type, and briefly describe the circumstances.

Thermal:

_____

_____

_____

_____

Mechanical:

_____

_____

_____

_____

Poisonous:

_____

_____

_____

_____

Corrosive:

_____

_____

_____

_____

Asphyxiation:

_____

_____

_____

_____

Radiation:

_____

_____

_____

_____

Etiological:

_____

_____

_____

2. Review the poison lines presented in the *Health and Safety* chapter. Sketch a poison line for hydrogen sulfide (H_2S). You will need to consult outside resource materials, such as SDSs for H_2S, to complete this exercise.

_____

_____

_____

_____

3. Compare your department's procedures for preventing or reducing heat stress to those listed in this chapter. Are there any procedures listed in the chapter that are not practiced in your department? Should they be added? Do you have suggestions for adding to the list included in the *Health and Safety* chapter?

_____

_____

_____

_____

4. Using outside reference materials, including the EPA and AIHA web sites, fill in the values for the following three hazardous chemicals:

Chemical	ERGP-2	1/10 IDLH	IDLH	TLV/TWA	TLV/STEL
Ammonia					
Chlorine					
Sulfur Dioxide					

Chemical	AEGL-2					
	5 min	10 min	30 min	60 min.	4 hr	8 hr
Ammonia						
Chlorine						
Sulfur Dioxide						

Important Terminology

The following are all important terms that you should know.

Parts per million (ppm)
Parts per billion (ppb)
Lethal dose (LD_{50})
Lethal concentration (LC_{50})
Permissible exposure limit (PEL)
Threshold limit value/time-weighted average (TLV/TWA)
Threshold limit value/ceiling (TLV/C)
Threshold limit value/short-term exposure limit (TLV/STEL)
Immediately dangerous to life or health (IDLH)
Emergency Response Planning Guideline (ERPG)
Acute exposure guideline levels (AEGL)
Radiation absorbed dose (RAD or rad)

On the line next to each term in Column A, print the letter of its definition from Column B.

Column A

Terminology

_____ **1.** Parts per million (ppm)

_____ **2.** Parts per billion (ppb)

_____ **3.** Lethal dose (LD_{50})

_____ **4.** Lethal concentration (LC_{50})

_____ **5.** Permissible exposure limit (PEL)

_____ **6.** Threshold limit value/time-weighted average (TLV/TWA)

_____ **7.** Threshold limit value/ceiling (TLV/C)

Column B

Definitions

A. An atmospheric concentration of any toxic, corrosive, or asphyxiant substance that poses an immediate threat to life, or would cause irreversible or delayed adverse health effects, or would interfere with an individual's ability to escape from a dangerous atmosphere.

B. The concentration of an inhaled substance that results in the death of 50% of the test population in a specific time period (usually 1 hour).

C. Numerically equivalent to 0.000,001 (10^{-6}).

D. The 15-minute, time-weighted average exposure that should not be exceeded at any time, nor repeated more than four times daily with a 60-minute rest period required between each STEL exposure.

E. The maximum concentration that should not be exceeded, even instantaneously.

F. Numerically equivalent to 0.000,000,001 (10^{-9}).

G. The maximum airborne concentration below which it is believed that nearly all individuals could be exposed for up to 1 hour without experiencing or developing irreversible or serious health effects or symptoms that could impair an individual's ability to take protective action.

	Column A		Column B
	Terminology		**Definitions**

_____ **8.** Threshold limit value/short-term exposure limit (TLV/STEL)

H. The maximum time-weighted concentration at which 95% of exposed, healthy adults suffer no adverse effects over a 40-hour work week.

_____ **9.** Immediately dangerous to life or health (IDLH)

I. Developed by the National Research Council's Committee on Toxicology to provide uniform exposure guidelines for the general public.

_____ **10.** *Emergency Response Planning Guideline* (ERPG-2)

J. Unit for radiation dose.

_____ **11.** Acute exposure guideline levels (AEGL)

K. The concentration of an ingested, absorbed, or injected substance that results in the death of 50% of the test population.

_____ **12.** Radiation absorbed dose (RAD or rad)

L. The maximum airborne concentration of a material to which an average healthy person may be exposed repeatedly for 8 hours each day, 40 hours per week, without suffering adverse effects.

Study Group Activity

1. As a group, examine and debate the "safety truths" presented in the chapter.

_____

_____

_____

_____

Study Group Learning Through Inquiry Scenario 2–1

You are the health and safety supervisor for a large chemical manufacturing plant with 1,500 employees. You have been assigned the task of developing a medical surveillance program for the members of the plant's Emergency Response Team (ERT). Facility hazards include flammable liquids and gases, corrosives, and poisonous liquids.

The volunteer ERT is trained to the Hazardous Materials Technician-level skills and includes personnel from the Operations, Maintenance, Safety, and Administration Departments within the chemical plant's workforce. In addition, there are two Maintenance Department members on each shift who are also trained to the Hazmat Technician level.

Given the information provided in the *Health and Safety* chapter, what would you recommend as the components of the medical surveillance program? Management has also asked for your recommendations on several specific issues:

1. Which criteria should be used for initial entry and reentry medical monitoring?

2. Because medical emergencies are handled by the off-site fire department, how can medical monitoring operations be performed?

3. Should the facility contact the local hospital? If yes, which type of information should be exchanged?

4. Several managers have stated that physical examinations can be provided on a 3-year interval. Do you agree or disagree with this statement? What is your response?

Study Group Learning Through Inquiry Scenario 2–2

You are the supervisor of a Hazardous Materials Response Team (HMRT) for a metropolitan fire department. One of your HMRT members informs you that she is pregnant and wants to continue working as long as possible up until the time that she will go on light-duty status. However, she questions you concerning the HMRT's policy regarding potential exposure to embryotoxins. Specifically, the team member wants your advice concerning how long it would be safe to work on the HMRT while pregnant. Your department's Policy and Procedures Manual provides clear guidance for fire fighters but does not address the issue for HMRT members.

After discussing the issue with the employee, you raise the issue with your fire chief. The chief agrees that a policy should be developed and asks you to consult with the department's medical officer. The chief wants you to address the following questions in a briefing paper for a future staff meeting. Use information provided in the *Health and Safety* chapter to develop your own opinion and also to gain more information regarding embryotoxins.

1. From a fire department policy perspective, should HMRT members be placed in a greater risk category than the average fire fighter? The chief agrees that fire fighters are in a high-risk occupation as compared to other occupations, but considering the special equipment, training, and procedures the HMRT uses, are HMRT members really at greater chronic health risk than any other employee who is not on the HMRT?

2. If you believe that the health risks of HMRT members are greater than risks of the average fire fighter, which key points will you work into the briefing paper the chief has requested? How will you support your case?

3. If HMRT members are really in a high-risk occupation, are female HMRT members exposed to an even greater risk while they are pregnant? Why or why not?

Summary and Review

1. What is the health hazard equation?

2. Identify the five routes of exposure:

3. Exposure to various types and doses of hazmats over a period of years is associated with chronic health effects. Identify at least two ways to monitor such exposures:

4. Identify three ways in which the human eye can be exposed via skin absorption:

5. What is the dose equation?

6. When evaluating the establishment of hazard control zones at hazmat emergencies, which of the following are generally the most informative: TLV, IDLH, ERPG, AEGL, or one-tenth the IDLH?

7. Carcinogens do not have an IDLH value, and many do not have a TLV value. Why?

8. List the four types of ionizing radiation:

9. Although sometimes found at remediation operations, air-cooled jackets and suits are typically not used for emergency response applications. Why?

10. The following statements about ice-cooled vests are all true except one. Which is false?
  a. Ice-cooled vests are a passive cooling system that operates on the principle of conductive heat cooling.
  b. Ice-cooled vests are relatively inexpensive and lightweight.
  c. Ice-cooled vests are not as effective as air-cooled units and water-cooled jackets.
  d. Ice-cooled vests can also be used with heat packs for operations in extremely cold working environments.

11. Which steps can be taken to prevent or minimize injuries from cold exposures?

12. There are two primary objectives of a medical surveillance program. The first is to determine that an individual can perform his or her assigned duties, including the use of personal protective clothing and equipment. What is the other objective?

13. What is the purpose of including nonemergency treatment in a medical surveillance program?

14. Identify at least three elements of a personal protective equipment (PPE) program:

15. Identify at least three components of a site safety plan:

16. Identify at least three topics that should be covered in a pre-entry safety briefing:

Self-Evaluation

Review all your work in this lesson and note your stronger and weaker areas.

Overall I feel I did (very well / well / fair / not so well) on the acronyms and abbreviations exercise.

Overall I feel I did (very well / well / fair / not so well) on the self-test questions.

Overall I feel did (very well / well / fair / not so well) on the terminology (matching) exercise.

Overall I feel I did (very well / well / fair / not so well) on the summary and review questions.

When compared to the previous lesson, I think I performed (better / worse / equally well).
List two areas in this chapter in which you feel you could improve your skill or knowledge level:

1. _____
   _____
   _____
   _____

2. _____
   _____
   _____
   _____

Consider the following self-evaluation questions as they pertain to this chapter:

Am I taking effective notes?

Am I dedicating enough quality time to my studies?

Is anything distracting my focus?

Was any part of this chapter too advanced for me?

Did I find that I don't have enough background experience to sufficiently grasp certain subject areas?

For areas in which I did particularly well, was it because I'm particularly interested in that subject matter? How so?

Were some things easier to learn because I have prior experience in learning or working with the concepts or principles?

Did I find that certain portions of the textbook seem to be better organized and effective in explaining key points?

Managing the Incident: Problems, Pitfalls, and Solutions

Chapter Orientation

Open the textbook to the *Managing the Incident: Problems, Pitfalls, and Solutions* chapter. When you have finished looking through the chapter, respond to the following items. Use the textbook as you jot down your comments in the spaces provided below.

1. Federal regulations require that both public safety and industrial emergency response organizations use a nationally recognized incident command system for emergencies involving hazardous materials. Beyond these regulatory requirements, why use ICS?

   _____

   _____

   _____

   _____

   _____

2. Reflect on your current level of knowledge or background experience about the topics covered in this chapter. Where have you read about, learned about, or applied this knowledge in the past?

   _____

   _____

   _____

   _____

   _____

3. Which sections or parts of this chapter strike you as looking especially interesting?

   _____

   _____

   _____

   _____

4. Which particular subjects in this chapter are important for a person in your position to master?

   _____

   _____

   _____

   _____

5. What do you predict will be the most difficult things for you to learn in this chapter?

_____

_____

_____

_____

_____

Learning Objectives

Examine the *Managing the Incident: Problems, Pitfalls, and Solutions* chapter objectives and respond to the following question:

1. Which objectives in this chapter do you feel you can achieve right now, with a reasonable level of confidence?

_____

_____

_____

_____

_____

As you read the sections of the chapter that deal with the objectives you have identified above, make sure your ideas and knowledge base match those of the authors. If they do not match, you should examine how your current understanding of the material differs from that of the authors. Depending on the level at which you wish to master the subject, discrepancies will have to be rectified and gaps will need to be filled.

Abbreviations and Acronyms

The following abbreviations and acronyms are used in the *Managing the Incident: Problems, Pitfalls, and Solutions* chapter:

ASO	Assistant Safety Officer
CBRNE	Chemical, Biological, Radiological, Nuclear, and High-Yield Explosive
CERFP	CBRNE Enhanced Response Force Package
CHEMTREC®	Chemical Transportation Emergency Center
CRM	Crew Resource Management
CST	Civil Support Team
DECON	Decontamination
DEQ	Department of Environmental Quality
DHS	Department of Homeland Security
DMORT	Disaster Mortuary Operational Response Team

EOC	Emergency Operations Center
EOD	Explosive Ordinance Disposal
ERP	Emergency Response Plan
ERT	Emergency Response Team
FSRT	Fatality Search and Recovery Team
HMRT	Hazardous Materials Response Team
HRF	Homeland Response Force
HSPD	Homeland Security Presidential Directive
IAP	Incident Action Plan
ICP	Incident Command Post
ICS	Incident Command System
ISO	Incident Safety Officer
IT	Information Technology
JIC	Joint Information Center
LEPC	Local Emergency Planning Committee
LNO	Liaison Officer
NIMS	National Incident Management System
NOAA	National Oceanic and Aeronautical Administration
NOP	Next Operational Period
OIM	Off-Shore Installation Manager
PACE Model	Primary Plan, Alternate Plan, Contingency Plan, Emergency Plan
PIO	Public Information Officer
PSAP	Public Safety Answering Point
R&D	Research and Development
RIT	Rapid Intervention Team
ROE	Rules of Engagement
RP	Responsible Party
SDS	Safety Data Sheets
SOP	Standard Operating Procedure
SWAT	Special Weapons and Tactics
UC	Unified Command; Unified Commander
US&R	Urban Search and Rescue
WMD	Weapons of Mass Destruction

Abbreviations and Acronyms Exercise

1. Which of the following would most likely have extensive knowledge of offshore legislation, drilling, production, maintenance, engineering, and safety management?

   a. JIC

   b. HRF

   c. OIM

   d. CST

2. Which of the following is a member of the ICS Command Staff?

    a. UC

    b. PIO

    c. RIT

    d. RP

3. Which of the following is deployed for the rescue of victims of structural collapse?

    a. SWAT

    b. ERT

    c. CST

    d. US&R

4. Which of the following would be most interested *primarily* in the environmental impact of a release?

    a. CHEMTREC®

    b. DHS

    c. DEQ

    d. HRF

For each of the following sentences, write in the correct abbreviation or acronym (from the preceding list) so that that sentence makes sense. Use each abbreviation or acronym only once.

5. _____ was originally defined in 1977 by aviation psychologist Dr. John Lauber as "using all available resources (information, equipment and people) to achieve safe and efficient flight operations." Key components include command, leadership, and resource management.

6. The _____ consists of the strategic goals, tactical objectives, and support requirements for the incident.

7. _____ is an organized system of roles, responsibilities, and standard operating procedures used to manage and direct emergency operations.

8. _____ is the baseline incident management system established under HSPD-5.

9. _____ are command-level representatives from each of the primary responding agencies who present their agency's interests as a member of a unified command team.

10. The organization legally responsible under government environmental laws for the clean-up of a hazmat release is called the _____.

Study Session Overview

Use the textbook to help you answer the questions or work from memory.

1. Which roles do law enforcement officers play in the ICS organization?

_____

_____

_____

_____

2. Where would you go to locate technical information specialists?

_____

_____

_____

_____

3. The textbook differentiates between an incident and a crisis. Give an example of an incident that happened in your community and briefly note how this incident could have transitioned to a crisis situation.

_____

_____

_____

_____

4. Review the resource management lessons presented in the chapter. Which one really stands out as true in your experience? In what way?

_____

_____

_____

_____

5. What does "command presence" mean to you?

_____

_____

_____

_____

6. From your experience, which factors contribute to the "reluctance of on-scene personnel to provide regular and timely updates to the ICP or EOC."

_____

_____

_____

_____

7. The chapter states that that an emergency can have a favorable technical or operational outcome and still be a political disaster. Briefly describe a hazmat emergency that you responded to or know about that resulted in substantial political problems, even though the overall outcome was operationally successful. Which factors contributed to the political problems?

_____

_____

_____

_____

8. Briefly outline the "rule of threes" for minimizing political vulnerability. How could the rule of threes have helped to minimize the political problem(s) associated with the incident you described above in question 7?

_____

_____

_____

_____

Self-Test

Use the textbook to help you answer the questions or work from memory.

Column A lists the players and participants who will interact within the ICS organization. On the line next to each term in Column A, print the letter of its definition from Column B.

Column A	Column B
Players	**Definitions**
_____ 1. Incident commander	**A.** Individuals who provide specific expertise to the IC either in person, by telephone, or through other means.
_____ 2. Unified commanders	**B.** Highly trained and equipped response teams who deliver a highly specialized response service and capability (e.g., urban search and rescue teams, bomb squads).
_____ 3. Investigators	**C.** Organization legally responsible under government law for the clean-up of a hazmat release.
_____ 4. Government officials	**D.** The individual responsible for establishing and managing the overall incident action plan (IAP).
_____ 5. Law enforcement officers	**E.** Individuals who receive "911" calls for assistance and dispatch appropriate units to the incident locations.
_____ 6. Emergency Response Teams (ERTs)	**F.** Individuals who normally do not have an on-scene emergency response function, but who are key players within the plant environment.
_____ 7. Special operations teams	**G.** Individuals who provide important support services at the incident (e.g., water and utility company employees, heavy equipment operators).
_____ 8. Communications personnel	**H.** Individuals who normally do not have an emergency response function but who bring a lot of political clout to the incident (e.g., mayors, city/county managers).
_____ 9. Responsible party	**I.** Crews of specially trained personnel used within business and industrial facilities for the control and mitigation of emergency situations.
_____ 10. Facility managers	**J.** Individuals who may provide both mitigation and support services at the incident (e.g., spill control, product transfer operations, site clean-up and recovery).
_____ 11. Support personnel	**K.** Individuals who are responsible for determining the origin and cause of the hazmat release, including any related evidence collection and preservation.
_____ 12. Technical information specialists	**L.** Command-level representatives from each of the primary responding agencies who present their agency's interests as a member of a unified command organization.
_____ 13. Environmental clean-up contractors	**M.** Resources for ensuring scene safety (i.e., scene and traffic control), perpetrator arrest or control, and evidence preservation.

14. Which of the following requires that both public safety and industrial emergency response organizations use a nationally recognized Incident Command System for hazmat emergencies?

    a. HSPD-8

    b. OSHA 1910.120(q)

    c. 40 CFR Part 68

    d. 29 CFR 1910.119

15. The National Incident Management System (NIMS) is a baseline incident management organization that is utilized by:

    a. the federal government.

    b. state and local governments.

    c. many private sector organizations.

    d. All of the above

16. What does "unity of command" mean?

    a. Work is assigned based on the functions to be performed

    b. Lines of authority are clearly defined

    c. Every person reports to only one supervisor

    d. No more than five individuals report to any one supervisor

17. At the very least, a(n) _____ must be identified on all incidents, regardless of their size.

    a. IC

    b. Safety Officer

    c. Liaison Officer

    d. Staging Officer

Column A lists the four sections with ICS. Match each section to its respective primary responsibility in Column B.

Column A	Column B
ICS Section	**Responsibility**
_____ 18. Operations Section	**A.** Conducts assessments and identifies the future needs, and then develops the plans required to support the response.
_____ 19. Planning Section	**B.** Responsible for getting funds where they are needed.
_____ 20. Logistics Section	**C.** Delivers the required tactical-level services in the field to make the problem go away, including fire, hazmat, emergency medical, and other services.
_____ 21. Administration/Finance Section	**D.** Provides all incident support needs, including facilities, services, and materials.

22. What do the Safety Officer, Liaison Officer, and Information Officer all have in common?

    a. They are all Command Staff officers.

    b. They all report directly to the hazmat group safety officer.

    c. They are all required by federal law to be staffed at all Level III hazmat incidents.

    d. They all report to and operate from the Emergency Operations Center (EOC).

23. The Service Branch and the Support Branch are both elements of which section?

    a. Operations Section

    b. Planning Section

    c. Logistics Section

    d. Administration/Finance Section

24. Staging is an element within which section?

   a. Operations Section

   b. Planning Section

   c. Logistics Section

   d. Administration/Finance Section

25. Whenever a situation is encountered that could immediately cause or has caused injuries to emergency response personnel, the term _____ should precede the radio transmission.

   a. "safety alert"

   b. "emergency traffic"

   c. "distress hail"

   d. "code urgent"

26. The Hazardous Materials Group Safety Officer is responsible for:

   a. establishing tactical objectives for the Hazardous Materials Entry Team.

   b. determining the appropriate level of decontamination to be provided.

   c. ensuring that health exposure logs and records are maintained for all Hazardous Materials Group personnel, as needed.

   d. ensuring that all hot zone operations are coordinated with the Operations Section chief or IC to ensure tactical goals are being met.

27. Which of the following is responsible for establishing a safe refuge area, if one is established?

   a. Entry/backup function

   b. Decontamination function

   c. Site access control function

   d. Medical function

Practice

1. Locate the course entitled "IS 700.a–National Incident Management System (NIMS), An Introduction" on the FEMA web site (http://training.fema.gov/EMIWeb/is/is700a.asp). This independent study course explains the purpose, principles, key components, and benefits of NIMS. Work through the course by yourself or with a study partner.

2. Starting with the equipment list provided in the *Managing the Incident: Problems, Pitfalls, and Solutions* chapter and using outside resources, develop a proposal to equip your department with a state-of-the-art incident command post (ICP) kit. You do not need to research costs, but you should be able to substantiate the need for the equipment or services you are proposing to acquire.

3. Develop a presentation that could be delivered to local officials that supports the need for an alternative EOC location. Use local examples to support your position, if possible.

Study Group Activity

1. One member of your group should briefly describe an incident from his or her own personal experience in which the ICS was not implemented at an incident, to the detriment of the outcome. Then as a group, everyone should propose ways in which the identified problems could have been minimized or avoided through a properly implemented ICS.

2. Divide into two groups. One group will develop the pros and cons of "people-dependent" response programs, while the other group develops the pros and cons of "system-dependent" response programs. The groups should rejoin and discuss their findings.

3. As a group, consider and discuss your reactions to the statement in the chapter: "In those cases where ICS has not resulted in the operational improvements expected, the problems are typically associated with planning, training, and the organization buying into the ICS program, as compared to the ICS system itself." Each member of the group should provide examples to support or refute this statement.

4. As a group, consider and discuss your reactions to the statement in the chapter: "There are no experts, but only information sources!" Each member of the group should provide examples to support or refute this statement.

Study Group Learning Through Inquiry Scenario 3–1

You are the emergency response supervisor within a petrochemical manufacturing facility (e.g., chemical plant, refinery, petrochemical plant). A series of incidents prompted the Corporate Safety Office to commission a third party safety audit of your facility. The following issues have been identified by the audit team:

- Confidential interviews with managers, supervisors, and operators have revealed strong evidence that the chain of command during an emergency is unclear.
- Interviews with Emergency Response Team members revealed that numerous supervisors and managers who have no emergency response training have been consistently reporting to the emergency scene. In many cases, these individuals do not wear appropriate levels of personal protective clothing.
- All emergency communications are maintained on a single emergency radio channel. This has led to numerous communications problems.
- The plant Emergency Operations Center (EOC) is maintained within the process area in a control room.
- Individuals are unclear as to their roles and responsibilities.

Your Plant Manager has asked you to give a presentation to the facility management team with your recommendations on how to resolve the problems identified by the audit. Based on the information provided, answer the following questions:

1. The vice president of the company plans to attend the meeting. As a safety professional, you see this as a golden opportunity to implement much needed changes to your emergency response program. You have been given 30 minutes on the meeting agenda to give your presentation. What are the three main recommendations that you want to make? (Review the materials on incident command in this chapter.)

2. How will you handle your presentation to the management team knowing that many of the managers who will be present at the meeting are the same people who routinely report to the emergency scene without the proper training or equipment? You know they are part of the problem. (Review ICS lessons learned in this chapter.)

Study Group Learning Through Inquiry Scenario 3–2

You are the supervisor of a newly formed Hazardous Materials Response Team, which is assigned to the Special Operations Division of the fire department. Your department has more than 800 employees assigned to 22 stations. Your department uses the incident command system very effectively and appoints a safety officer on all working incidents. The new HMRT has been integrated smoothly into the command structure; however, you have had some significant problems distinguishing the difference between the incident safety officer and the hazardous materials group safety officer (i.e., Assistant Safety Officer–Hazmat). You are concerned about establishing the credibility of the HMRT within the command structure of the department.

You raise the issue at an Operations Division staff meeting and the operations chief asks you, "Why do we need two safety officers at the same incident?" Based on the information in the *Managing the Incident: Problems, Pitfalls, and Solutions* chapter, answer the following questions:

1. Are two safety officers really necessary at hazardous materials incidents? If you feel that the ASO-Hazmat is necessary, how would you answer the chief's question and justify your position?

2. From a functional perspective, which duties and responsibilities should the incident safety officer perform as compared to the ASO-Hazmat?

3. From a political perspective, what can you do within the HMRT to assure that the system runs better?

Summary and Review

1. ICS is predicated on basic management concepts, including division of labor and clearly defined lines of authority. Identify at least one other management concept:

2. Which of the following is the ICS organizational level that has functional responsibility for primary functions of emergency incident operation?
   a. Sections
   b. Branches
   c. Divisions/groups
   d. Command staff

3. The ICS organizational structure develops in a _____ fashion based on the size and nature of the incident.
   a. modular
   b. linear
   c. hierarchal
   d. pyramid

4. When the ICP and the EOC are both operating simultaneously at a major incident, the ICP is primarily oriented toward tactical control issues pertaining to the on-scene response. What does the EOC deal with?

5. List at least four equipment items that should be located at the EOC:

6. True / False: Communications of a sensitive nature should not be given over nonsecure cellular telephones or radios that can be monitored.

7. True / False: Unified command is not management by committee; there will always be a lead agency or one agency that has 51% of the vote as compared to the other players.

8. What is the most effective way to ensure that a consolidated plan of action is implemented?

Column A lists the five primary functions and two secondary support functions assigned to the Hazardous Materials Group. Match each function to its respective tasks/responsibilities in Column B.

Column A	Column B
Hazardous Materials Group Function	**Tasks/Responsibilities**

_____ **9.** Safety function

    A. Establishes hazard control zones, establishes and monitors egress routes at the incident site, and ensures that contaminants are not being spread.

_____ **10.** Entry/backup function

    B. Responsible for pre- and post-entry medical monitoring and evaluation of all entry personnel, and provides technical medical guidance to the Hazardous Materials Group, as requested.

_____ **11.** Decontamination function

    C. Responsible for all entry and backup operations within the hot zone, including reconnaissance, monitoring, sampling, and mitigation.

_____ **12.** Site access control function

    D. Responsible for ensuring that safe and accepted practices and procedures are followed throughout the course of the incident.

_____ **13.** Information/research function

    E. Responsible for control and tracking of all supplies and equipment used by the Hazardous Materials Group during the course of an emergency.

_____ **14.** Medical function

    F. Responsible for gathering, compiling, coordinating, and disseminating all data and information relative to the incident.

_____ **15.** Resource function

    G. Responsible for the research and development of the decontamination plan, setup, and operation of an effective decontamination area capable of handling all potential exposures, including entry personnel, contaminated victims, and equipment.

16. The Hazardous Materials Group supervisor will usually report to either the incident commander (IC) or the:

  a. Operations Section Chief.

  b. Hazardous Materials Resource Leader.

  c. safety officer.

  d. Hazardous Materials Branch Director.

17. Which of the following Hazardous Materials Group staff is responsible for directing rescue operations within the hot zone?

  a. Decontamination Team

  b. Entry Team

  c. Site access control

  d. Hazardous Materials Medical Unit

18. Which of the following Hazardous Materials Group staff is responsible for providing recommendations for the selection and use of protective clothing and equipment?

  a. Decontamination Team

  b. Hazardous Materials Medical Unit

  c. Hazardous Materials Group Safety Officer

  d. Hazardous Materials Information/Research Team

19. In the heat of battle, the Operations Section may start to reduce the flow of information to the remainder of the ICS organization. Identify at least one way to avoid this problem:

20. How can the Liaison Officer help manage the political issues of a hazmat incident?

Self-Evaluation

Review all your work in this lesson and note your stronger and weaker areas.

Overall I feel I did (very well / well / fair / not so well) on the acronyms and abbreviations exercise.

Overall I feel I did (very well / well / fair / not so well) on the self-test questions.

Overall I feel I did (very well / well / fair / not so well) on the summary and review questions.

When compared to the previous lessons, I think I performed (better / worse / equally well).

List two areas in this chapter in which you feel you could improve your skill or knowledge level:

1. _____
   _____
   _____
   _____

2. _____
   _____
   _____
   _____

Consider the following self-evaluation questions as they pertain to this chapter:

Am I taking effective notes?

Am I dedicating enough quality time to my studies?

Is anything distracting my focus?

Was any part of this chapter too advanced for me?

Did I find that I don't have enough background experience to sufficiently grasp certain subject areas?

For areas in which I did particularly well, was it because I'm particularly interested in that subject matter? How so?

Were some things easier to learn because I have prior experience in learning or working with the concepts or principles?

Did I find that certain portions of the textbook seem to be better organized and effective in explaining key points?

The Eight Step Process©: An Overview

Chapter Orientation

Open the textbook to *The Eight Step Process©: An Overview* chapter. Keep in mind that this chapter is designed as a "bridge chapter" that provides an overview of the Eight Step Process©, which is a systematic way of approaching a hazmat incident. You will learn about each step in the process in greater detail as you progress through the remaining chapters in the textbook. When you have finished looking through the chapter, respond to the following items. Use the textbook as you jot down your comments in the spaces provided below.

1. In general terms, how does the use of standardized procedures, such as the Eight Step Process© affect safety?

   _____
   _____
   _____
   _____

2. Reflect on your current level of knowledge or background experience regarding the Eight Step Process© as covered in this chapter. Where have you read about, learned about, or applied this knowledge in the past?

   _____
   _____
   _____
   _____

3. Which sections or parts of this chapter strike you as looking especially interesting?

   _____
   _____
   _____
   _____

4. Which particular subjects in this chapter are important for a person in your position to master?

   _____
   _____
   _____
   _____

5. What do you predict will be the most difficult things for you to learn in this chapter?

   _____
   _____
   _____
   _____

Learning Objectives

Examine *The Eight Step Process©: An Overview* chapter objectives and respond to the following question:

1. Which objectives in this chapter do you feel you can achieve right now, with a reasonable level of confidence?

_____

_____

_____

_____

_____

As you read the sections of the chapter that deal with the objectives you have identified above, make sure your ideas and knowledge base match those of the authors. If they do not match, you should examine how your current understanding of the material differs from that of the authors. Depending on the level at which you wish to master the subject, discrepancies will have to be rectified and gaps will need to be filled.

Abbreviations and Acronyms

The following abbreviations and acronyms are used in *The Eight Step Process©: An Overview* chapter:

APR	Air-Purifying Respirator
DECON	Decontamination
EOC	Emergency Operations Center
IAP	Incident Action Plan
IC	Incident Commander
ICP	Incident Command Post
IED	Improvised Explosive Devices
LPG	Liquefied Petroleum Gas
PASS	Personal Alert Safety System
PERO	Post-Emergency Response Operations
PPA	Public Protective Actions
PPE	Personal Protective Equipment
PPV	Positive Pressure Ventilation
SBCCOM	U.S. Army Soldiers Biological and Chemical Command
SWAT	Special Weapons and Tactics Team

Abbreviations and Acronyms Word Search Puzzle

Locate 14 of the 15 abbreviations and acronyms (from the preceding list) in the word puzzle below.

```
L  U  S  K  R  U  A  M  N  R  I  V  Q
V  U  W  G  D  J  B  M  P  R  S  O  L
C  B  F  E  U  H  E  F  R  B  A  O  P
E  Z  I  K  A  Y  R  T  C  P  D  E  G
J  P  C  I  U  F  M  C  T  X  A  K  M
E  B  A  E  R  X  O  A  C  Z  B  X  U
P  P  A  S  S  M  R  V  Y  D  J  P  P
P  H  H  X  M  E  E  A  Z  N  V  P  N
P  E  V  S  M  T  P  T  C  P  B  V  T
U  P  O  E  W  Q  H  O  D  I  D  E  S
Y  M  A  C  K  A  A  S  X  K  I  B  H
Y  J  M  Q  G  F  T  V  S  Q  F  O  J
Z  I  M  J  J  M  F  D  B  H  C  R  T
```

Study Session Overview

Use the textbook to help you answer the questions or work from memory.

1. The eight functions in the Eight Step Process© typically follow an implementation timeline at the incident. Based on your current understanding of the eight steps, which one is most likely to be ongoing throughout the entire incident until it is terminated?

_____

_____

_____

_____

2. Which of the Eight Step Process© functions is considered the foundation on which all subsequent response functions and tactics are built?

_____

_____

_____

_____

3. Which comes first: conducting hazard and risk evaluation or selecting personal protective clothing and equipment?

_____

_____

_____

_____

_____

4. Which comes first: committing personnel to the hot zone or evaluating the hazards and risks?

_____

_____

_____

_____

_____

5. Which comes first: developing an incident action plan or evaluating the hazards and risks?

_____

_____

_____

_____

_____

6. What would be your primary concerns when conducting mass decontamination operations at an incident involving possible terrorism? How would you address those concerns?

_____

_____

_____

_____

_____

7. In your own words, why is it important to formally terminate the incident?

_____

_____

_____

_____

_____

Self-Test

Use the textbook to help you answer the questions or work from memory.

Match each of the steps in the Eight Step Process© (Column A) to its respective goal in Column B.

Column A

Eight Step Process©

Column B

Goal

_____ **1.** Site Management and Control

A. To ensure that all emergency response personnel have the appropriate level of personal protective clothing and equipment for the expected tasks.

_____ **2.** Identify the Problem

B. To ensure that the incident priorities (i.e., rescue, incident stabilization, environmental and property protection) are accomplished in a safe, timely, and effective manner.

_____ **3.** Hazard and Risk Evaluation

C. To ensure the safety of both emergency responders and the public by reducing the level of contamination on scene and minimizing the potential for secondary contamination beyond the incident scene.

_____ **4.** Select Personal Protective Clothing and Equipment

D. To establish the playing field so that all subsequent response operations can be implemented both safely and effectively.

_____ **5.** Information Management and Resource Coordination

E. To identify the scope and nature of the problem, including the type and nature of hazardous materials involved as appropriate.

_____ **6.** Implement Response Objectives

F. To provide for the timely and effective management, coordination, and dissemination of all pertinent data, information, and resources between all of the players.

_____ **7.** Decon and Clean-Up Operations

G. To ensure that overall command is transferred to the proper agency when the emergency is terminated and that all post-incident administrative activities are completed per local policies and procedures.

_____ **8.** Terminate the Incident

H. To assess the hazards present, evaluate the level of risk, and establish an incident action plan (IAP) to make the problem go away.

9. Which of the following functions establishes the playing field for the layers (responders) and the spectators (everyone else)?
   a. Site management and control
   b. Hazard and risk evaluation
   c. Information management and resource coordination
   d. Implement response objectives

10. Site management and control involves:
   a. establishing a hot zone.
   b. analyzing container shapes.
   c. consulting technical information specialists.
   d. selecting personal protective clothing.

11. What is the primary objective of the risk evaluation process?
   a. Identify the presence of improvised explosive devices (IEDs)
   b. Verify the hazardous materials involved in the incident
   c. Determine whether responders should intervene
   d. Initiate public protective actions

12. Which of the following functions cannot be effectively accomplished unless a unified ICS organization is in place?
    a. Site management and control
    b. Hazard and risk evaluation
    c. Select personal protective clothing and equipment
    d. Information management and resource coordination

13. Which of the following strategies is not a defensive mode response objective?
    a. Rescue
    b. Fire control
    c. Spill control
    d. Public protective actions

14. The termination of emergency response operations includes which of the following activities?
    a. Decontamination
    b. Clean-up operations
    c. Leak control
    d. Incident debriefing

Practice

Assume you are responding to a train wreck within a half-mile of a mixed residential and business neighborhood in your community. There are 14 derailed cars, two of which are carrying 90 tons of chlorine each. One of the tank cars has ruptured and is releasing chlorine gas to the atmosphere. Match each of the emergency procedures listed in Column B to one of the eight steps in Column A. One answer is already provided (see Step 3). There are two procedures that match each step.

Column A	Column B
Eight Step Process©	**Procedure**
Step 1: Site Management and Control _____ _____	**A.** Ensure proper decontamination of emergency personnel before they leave the scene. **B.** Assure that site emergency workers are using the proper protective equipment and clothing equal to the hazards present.
Step 2: Identify the Problem _____ _____	**C.** Evaluate the risks of personnel intervening directly in the emergency. **D.** Initiate offensive tactics that will reduce or stop the flow of chlorine if that task can be accomplished without undue risk.
Step 3: Hazard and Risk Evaluation <u>C</u> _____ _____	**E.** Confirm that your command post is in a safe area and that your position will not be overrun by the migrating vapor cloud. **F.** Determine the concentrations of toxic gases present using both fixed monitors (if available) and portable instruments. What is the concentration of chlorine?
Step 4: Select Personal Protective Clothing and Equipment _____ _____	**G.** Identify, confirm, and verify the problem. If multiple problems exist, prioritize them and make independent assignments. **H.** Use a massive rinse on the outer shell of protective clothing. Maintain respiratory protection throughout the decontamination process.

(Continued)

Column A	Column B
Eight Step Process©	**Procedure**

Step 5: Information Management and Resource Coordination

I. Obtain the names and telephone numbers of all key individuals. Include contractors, public officials, and members of the media.

_____

J. Restrict access to the emergency site to authorized essential personnel.

_____

Step 6: Implement Response Objectives

K. Determine the nature, extent, and potential impact of the release.

L. Establish a command post well outside the chlorine vapor cloud area.

_____

_____

Step 7: Decon and Clean-up Operations

M. Order specialized equipment and expertise early in the incident. If you are unsure what your requirements are, always call for the highest level of assistance available.

_____

N. Ensure that a properly equipped backup rescue team is in place before initiating offensive tactics.

_____

Step 8: Terminate the Incident

O. Document all equipment or supplies used during the incident.

_____

P. Coordinate your emergency plans with all support personnel. Make sure that they are aware of where the hot, warm, and cold zones are located and that special hazards are involved.

_____

Study Group Activity

1. As a group, review and discuss the "Responder Tips" for each of the Eight Step Process© functions. Identify those that you feel are particularly important or valid.

_____

_____

_____

_____

_____

2. Brainstorm ways in which security issues can be handled or anticipated using the principles of the Eight Step Process©. Be sure to consider all eight steps, in sequence.

_____

_____

_____

_____

_____

Study Group Learning Through Inquiry Scenario 4–1

You are a battalion fire chief assigned to lead a post-incident analysis team investigating a fire fighter fatality. The fire chief wants to know if the department's standard operating procedures were followed.

The incident involved a warehouse that stored 18,000 one-pound LPG cylinders in an unsprinklered building. When the first engine company arrived at 3:00 A.M., the fire was venting through the skylight in the center of the warehouse. The company officer ordered an interior fire attack using handlines. When the battalion fire chief arrived on scene, the propane cylinders began to fail, intensifying the fire. The engine company was ordered out of the building, but a partial roof collapse trapped one fire fighter.

1. Using the Eight Step Process© shown in this chapter, determine the step in which the first-arriving engine company was engaged when the battalion chief ordered the company from inside the warehouse.

2. Which step in the Eight Step Process© was the battalion fire chief assigned to during the post-incident analysis?

3. During your interview with the battalion fire chief who commanded the fire, he informs you that he believed that the engine company officer failed to consider the risk of the propane cylinders inside the warehouse and that this omission was the major contributing factor in the fire fighter fatality. Do you agree with this statement? If so, why? If you disagree, what is the basis for your opinion? (Review the "Step 3" material in this chapter and use it to support your opinion.)

Summary and Review

1. Eight basic functions must be evaluated at emergencies involving, or suspected of involving, hazardous materials or WMD agents. Several of these functions are provided below. Fill in the blanks with the remaining functions, in correct sequence:

   Step 1. Site Management and Control

   Step 2. _____

   Step 3. _____

   Step 4. Select Personal Protective Clothing and Equipment

   Step 5. _____

   Step 6. Implement Response Objectives

   Step 7. _____

   Step 8. Terminate the Incident

2. Which of the Eight Step Process© functions is concerned with conducting an incident debriefing session for on-scene response personnel?

3. Which of the Eight Step Process© functions is concerned with providing regular updates to the local Emergency Operations Center (EOC), if activated?

4. What is the minimum level of training at which personnel should be able to deliver emergency decon?

   a. First Responder at the Awareness Level

   b. First Responder at the Operations Level

   c. Hazardous Materials Technician

   d. Specialist Employee A

5. Why does the selection of personal protective clothing and equipment come after hazard and risk evaluation instead of ahead of hazard and risk evaluation?

6. Identify at least three primary objectives of Step 1: Site Management and Control:

7. You are the incident commander at a chemical plant fire. You have decided to implement nonintervention mode until additional personnel and equipment arrive. Identify one nonintervention tactic that should be initiated:

8. List three clues suggesting that an improvised explosive device (IED) might be present at an incident:

9. A security officer on patrol on the ground level of a hospital parking garage smells a strong odor of gas. The officer calls 911 and requests that the fire department respond. The fire department dispatches one engine company to investigate. You are the company officer on the response.

   a. Which other information would you want to know while en route?

   b. Identify three ways you could perform defensive recon:

   c. Identify two or three ways to perform offensive recon:

10. There are three critical success factors in hour 1 of a hazmat response. The first is the ability of the responders to recognize clues that the incident may involve hazardous materials. What are the other two factors?

Self-Evaluation

Review all your work in this lesson and note your stronger and weaker areas.

   Overall I feel I did (very well / well / fair / not so well) on the acronyms and abbreviations exercise.

   Overall I feel I did (very well / well / fair / not so well) on the self-test questions.

   Overall I feel did (very well / well / fair / not so well) on the practice exercise.

   Overall I feel I did (very well / well / fair / not so well) on the summary and review questions.

   When compared to the previous lessons, I think I performed (better / worse / equally well).

List two areas in this chapter in which you feel you could improve your skill or knowledge level:

1. _____

_____

_____

_____

2. _____

_____

_____

_____

Consider the following self-evaluation questions as they pertain to this chapter:

Am I taking effective notes?

Am I dedicating enough quality time to my studies?

Is anything distracting my focus?

Was any part of this chapter too advanced for me?

Did I find that I don't have enough background experience to sufficiently grasp certain subject areas?

For areas in which I did particularly well, was it because I'm particularly interested in that subject matter? How so?

Were some things easier to learn because I have prior experience in learning or working with the concepts or principles?

Did I find that certain portions of the textbook seem to be better organized and effective in explaining key points?

Site Management

Chapter Orientation

Open the textbook to the *Site Management* chapter. When you have finished looking through the chapter, respond to the following items. Use the textbook as you jot down your comments in the spaces provided below.

1. In general terms, what makes learning how to establish control of the incident scene so important?

   _____

   _____

   _____

   _____

   _____

2. Reflect on your current level of knowledge or background experience about site management and control. Where have you read about, learned about, or applied the principles of site management and control in the past?

   _____

   _____

   _____

   _____

   _____

3. Which sections or parts of this chapter strike you as looking especially interesting?

   _____

   _____

   _____

   _____

   _____

4. Which particular aspects of site management and control are important for a person in your position to master?

   _____

   _____

   _____

   _____

5. What do you predict will be the most difficult things for you to learn in this chapter?

   _____

   _____

   _____

   _____

Learning Objectives

Examine the *Site Management* chapter objectives and respond to the following question:

1. Which objectives in this chapter do you feel you can achieve right now, with a reasonable level of confidence?

_____

_____

_____

_____

_____

As you read the sections of the chapter that deal with the objectives you have identified above, make sure your ideas and knowledge base match those of the authors. If they do not match, you should examine how your current understanding of the material differs from that of the authors. Depending on the level at which you wish to master the subject, discrepancies will have to be rectified and gaps will need to be filled.

Abbreviations and Acronyms

The following abbreviations and acronyms are used in the *Site Management* chapter:

ACH	Air Changes per Hour
ATSDR	Agency for Toxic Substances Disease Registry
CTN	Critical Transportation Needs
EAS	Emergency Alerting System
EBA	Escape Breathing Apparatus
ERG	*Emergency Response Guidebook*
HVAC	Heating, Ventilating, and Air Conditioning
ICP	Incident Command Post
IDLH	Immediately Dangerous to Life and Health
NOAA	National Oceanographic and Atmospheric Administration
NWS	National Weather Service
PID	Photoionization Device
PLAN	Personalized Localized Alerting Network
PPA	Public Protective Actions
SBS	Sick Building Syndrome
STAM	Staging Area Manager
VOC	Volatile Organic Compounds

Abbreviations and Acronyms Exercise

For each of the following sentences, write in the correct abbreviation or acronym (from the preceding list) so that that sentence makes sense.

1. Sources of SBS problems include copy machines, cleaning agents, recently applied pesticides, and chemicals that may release _____.

2. Biological contaminants from inside the building may also cause _____ problems.

3. _____ devices typically have 5 to 10 minutes of breathing air in their cylinders.

4. If there is no published value, consider using an estimated _____ of 10 times the TLV/TWA.

5. Law enforcement plays a very important role in providing security for the _____ and emergency responders within the isolation perimeter.

6. Many locations have improved their alerting systems by providing detailed _____ instructions in the local telephone directory.

7. Special weather radios are available that are activated by the _____ for severe storm warnings.

8. Commercial motor coaches are the only practical and cost-effective way of moving large numbers of people with _____.

9. The _____ is an effective method of alerting people in buildings and automobiles.

Study Session Overview

Use the textbook to help you answer the questions or work from memory.

1. In your own words, which safety issues are associated with site management and control?

   _____

   _____

   _____

   _____

   _____

2. What can happen at a hazmat incident if there is no strong, centralized command?

   _____

   _____

   _____

   _____

   _____

3. In what ways do the lessons learned from the Houston, TX, ammonia incident help advance our tactical understanding of site management and control?

_____

_____

_____

_____

_____

4. Under which circumstances are law enforcement or security personnel the best people to use to provide perimeter control and security? Give several examples.

_____

_____

_____

_____

_____

Self-Test

Use the textbook to help you answer the questions or work from memory.

1. The major emphasis of site management is on establishing control of the incident scene and:
   a. gathering information.
   b. designating safety zones.
   c. isolating people from the problem.
   d. alerting the public.

2. The isolation perimeter is always the line between the:
   a. inner perimeter and the outer perimeter.
   b. general public and the cold zone.
   c. area of refuge and the hot zone.
   d. immediate site of the spill and staging area.

3. Site management consists of all of the following tasks _except_:
   a. assuming command.
   b. establishing staging.
   c. establishing hazard control zones.
   d. implementing defensive confinement tactics.

4. As an incident grows or escalates, the IC designates a fixed location (staging area) where resources responding beyond the initial response can be placed until given a tactical assignment. Historically, this strategy was often known as:
   a. Level I staging.
   b. Level II staging.
   c. exterior staging.
   d. base staging.

5. Staging procedures facilitate safety and:
   a. command presence.
   b. backup refuge.
   c. accountability.
   d. perimeter security.

6. The following statements about the isolation perimeter are all true except one. Which one is false?
   a. The isolation perimeter is set up to maintain safety and security.
   b. Designating the isolation perimeter is an incident command responsibility.
   c. Proper protective clothing and equipment must be worn when establishing the isolation perimeter.
   d. Isolation perimeters are not practical for indoor hazmat incidents.

7. What is the primary purpose of establishing three different hazard control zones?
   a. Assist initial size-up
   b. Allocate resources more efficiently
   c. Provide control and personnel accountability
   d. Reduce radio communications

8. The area of refuge should be established within the:
   a. hot zone.
   b. warm zone.
   c. cold zone.
   d. staging area.

9. Decisions regarding the size of hazard control zones should be based on measures of:
   a. flammability.
   b. toxicity.
   c. radioactivity.
   d. All of the above

10. The following points all apply to hazard control zones except one. Which one is *not* correct?
   a. Hazard control zones should be marked and posted on the IC's tactical command worksheet.
   b. Initial monitoring efforts should concentrate on determining if IDLH concentrations are present.
   c. Once established, hazard control zones should not change or be modified.
   d. Property owners should be briefed on how and why you have established hazard control zones.

11. Public protective actions (PPAs) are the strategy used by the Incident Commander to protect the general public from the hazmat by implementing:
   a. protection-in-place.
   b. evacuation.
   c. either protection-in-place *or* evacuation, but never both at the same time.
   d. protection-in-place, evacuation, or a combination of both.

12. Protection-in-place is usually the best option when any of the following conditions exist *except*:
   a. when explosive or reactive materials are involved.
   b. when leaks can be rapidly controlled at their source.
   c. when short-duration solid or liquid leaks are present.
   d. when the hazmat has been totally released from its container and is dissipating.

13. Which of the following factors most strongly influences how successful protection-in-place in a particular building will be?
    a. The age and construction of the building
    b. The location of the building
    c. The occupancy of the building
    d. The terrain surrounding the building

14. The following are all indicators of sick buildings *except*:
    a. building occupants complain of symptoms associated with acute discomfort, such as headache.
    b. building occupants complain of fatigue and/or difficulty in concentrating.
    c. the symptoms are caused by a known hazmat spill inside the building.
    d. most complainants report relief soon after leaving the building.

15. A study conducted by the Battelle Human Affairs Research Center indicated that the cost to the manufacturing sector for full-scale evacuation is approximately _____ the cost of protection-in-place.
    a. two times
    b. three times
    c. seven times
    d. fifteen times

16. When the decision is made to commit to a full-scale public evacuation, four critical issues must be addressed and managed effectively: 1) alerting and notification; 2) transportation; 3) relocation facility; and 4):
    a. monitoring.
    b. information.
    c. manpower.
    d. multi-agency coordination.

17. One of the most frequent alerting system problems encountered in fixed facilities (such as a refinery) is:
    a. not knowing where to go once the alarm sounds.
    b. hesitancy among workers to leave workstations unattended.
    c. accounting for the whereabouts of contractors and visitors.
    d. the confusion created by a single warning tone that may also be used to indicate the beginning or ending of a work shift.

18. A good community alerting and notification system is based on:
    a. low-power AM radio systems.
    b. sirens and alarms.
    c. door-to-door visits.
    d. a variety of warning systems.

19. One study suggests that approximately _____ of a town or city's population will require some form of public shelter during an evacuation.
    a. 5%
    b. 10%
    c. 35%
    d. 50%

20. Which of the following is the correct term for people who have no means of transportation during an evacuation?

   a. Critical transportation needs

   b. Essential transports

   c. Special evacuation customers

   d. Transportation-dependent persons

Practice

1. Assume that a threat has been received that a car bomb is located in a parking garage nearby your state capital. Using a large-scale map and a sketch of the garage, diagram your initial isolation perimeter and explain your rationale.

_____

_____

_____

_____

2. Develop a set of user-friendly instructions for your community that explains the concept of protection-in-place and provides clear instructions on what residents should do to shelter in their own homes.

_____

_____

_____

_____

3. Assume that you have to alert visitors and staff at a large regional park that a train carrying hazardous materials (unknown at this time) has derailed upwind of the park. Which of the alerting methods found in this chapter would you use? Explain the basis for your selection. You should assume that the park includes trails, primitive camping, a lodge, several vendors, a large recreational lake, a dam, and two roads that lead to an interstate highway.

_____

_____

_____

_____

4. Identify several potential relocation facilities in your town or city. Explain why you have chosen them and which elements need to be in place to assure they will be effective in the event of a large-scale evacuation.

_____

_____

_____

_____

Important Terminology

The following are all important terms that you should know.

Staging
Isolation perimeter
Hazard control zones
Hot zone
Warm zone
Cold zone
Area of refuge
Public protective actions (PPAs)
Protection-in-place
Evacuation

On the line next to each term in Column A, print the letter of its definition from Column B.

Column A

Terminology

Column B

Definitions

_____ **1.** Staging

_____ **2.** Isolation perimeter

_____ **3.** Hazard control zones

_____ **4.** Evacuation

_____ **5.** Protection-in-place

_____ **6.** Hot zone

_____ **7.** Warm zone

_____ **8.** Cold zone

_____ **9.** Area of refuge

_____ **10.** Public protective actions (PPAs)

A. The designated crowd control line surrounding the hazard control zones; always the line between the general public and the cold zone. Law enforcement personnel may also refer to this as the outer perimeter.

B. The control zone at a hazardous materials incident site where personnel and equipment decontamination and hot zone support take place.

C. Designation of areas at a hazardous materials incident based on safety and the degree of hazard.

D. A holding area within the hot zone where personnel are controlled until they can be safely decontaminated or treated.

E. The controlled relocation of people from an area of known danger or unacceptable risk to a safer area or one in which the risk is considered to be acceptable.

F. The strategy used by the incident commander to protect the general population from a hazardous material by implementing a strategy of either (1) protection-in-place, (2) evacuation, or (3) a combination of protection-in-place and evacuation.

G. The location where resources can be placed while awaiting a tactical assignment.

H. The control zone immediately surrounding a hazardous materials incident that extends far enough to prevent adverse effects from hazardous materials releases from reaching personnel outside the zone. Law enforcement personnel may also refer to this as the inner perimeter.

I. Directing people to go inside a building, sealing it up as effectively as possible, and remaining there until the danger from a hazardous materials release has passed.

J. The areas at a hazardous materials incident that contains the command post and other support functions necessary to control the incident.

Study Group Activity

1. Working individually or in pairs, members of the group should conduct neighborhood surveys throughout their jurisdiction to identify the types of buildings present with regard to protection-in-place potential. Other group members should identify seasonal prevailing wind directions and topography that could affect the movement of toxic gas. Report your findings back to the group.

2. Reread the "Voices of Experience" feature in the *Site Management* chapter, in which several incidents that had poor hazard control zones are described. In an open discussion and working from their own personal experiences, group members should provide examples of incidents that had a poor hazard control zone, explain the factors that contributed to this situation, and list the key lessons learned from each incident.

Study Group Learning Through Inquiry Question 5–1

You are a member of the Local Emergency Planning Committee (LEPC) within your community. You have been appointed chairperson of a subcommittee to review ideas for improving the system by which public protective actions are implemented; the subcommittee was formed as a result of a recent hazmat incident where some citizens had to be evacuated. Currently, the police department is responsible for implementing the evacuation order, school district buses are used for any transportation, and local schools and churches are used as evacuation centers. The agreements to provide these services have been long-standing but there is no formal arrangement that guarantees their availability. The community has no procedure for sheltering-in-place.

High-hazard areas that the LEPC has identified within the community include two anhydrous ammonia storage facilities, a water treatment plant, several farm chemical facilities, an LPG and flammable liquid storage facility, a four-lane state highway, and a major east/west railway corridor.

Based on the background information provided and the information provided in this chapter, respond to the following questions:

1. Based on what you know so far, are there any particular problems that you feel should be addressed by your subcommittee? Develop a prioritized list of your top three concerns.

2. Based on the hazards identified in the community, which ones will place the community at greatest risk in terms of the need for public protective actions?

3. In your opinion, does this level of risk justify the time it would take to sell members of the LEPC on the concept of sheltering-in-place or protection-in-place? If you believe that this effort is worth the time, how would you develop and support your position before the LEPC?

4. Assuming that the LEPC supports your proposal on the concept of sheltering-in-place or protection-in-place, outline a plan of action for selling the concept to the community.

Study Group Learning Through Inquiry Question 5–2

You are the Fire Chief in a small town with a volunteer fire department. Under your leadership, the department has organized and conducted a series of emergency response exercises that were designed to test its ability to alert the public during a hazmat emergency. A major railway and highway pass through your community.

The most recent exercise revealed some significant findings, including the following problems:

- The local 911 system cannot handle the influx of telephone calls that are expected during a major event such as a train derailment requiring evacuation. The local telephone company uses old equipment, but will not spend the money required to upgrade the system for an event that may never happen.
- Your town has one AM/FM radio station that is staffed as a "live radio station" during the day, but uses a taped unstaffed format from 8:00 P.M. to 5:00 A.M. The recent test of the Emergency Broadcast System revealed that the EBS interrupt connection was not functioning properly. In other words, the EBS could not be activated.

Based on the background information provided and the information discussed in this chapter, respond to the following questions:

1. How could you overcome the two problems identified during the exercise critique? Could you build an effective alerting and notification system? If you answered "yes," which major elements would you rely on to build your public protective action plan?

2. If your answer to question 1 is "no," how would you overcome these two significant problems?

3. If a real incident required an evacuation and several citizens were seriously injured during that process, how would the image of your fire department be affected if the two problems described previously were revealed in a follow-up investigation? What if you had previously identified these problems in an exercise but nothing was ever done about the problem? Would you be personally liable for failure to correct the problem? If you answered "no," why wouldn't you be liable?

Summary and Review

1. Define site management.

2. Six major tasks must be implemented as part of the site management and control process. Five are listed below. Which task is missing from this list?
   a. Assuming command and establishing control of the incident scene
   b. Assuring safe approach and positioning of emergency response resources at the incident scene
   c. Establishing staging as a method of controlling arriving resources
   d. Establishing hazard control zones to assure a safe work area for emergency responders and supporting resources
   e. Sizing up the need for immediate rescue and implementing initial public protective actions

3. There are two general guidelines for the safe approach and position of emergency response personal at a hazmat incident. One is to approach from uphill and upwind whenever possible. What is the other?

4. The staging area _____ is responsible for managing all activities within a staging area.

    a. manager

    b. officer

    c. chief

    d. director

5. Who is usually responsible for maintaining the isolation perimeter throughout the incident?

    a. The staging officer

    b. Law enforcement or security professionals

    c. First-arriving emergency response units

    d. Emergency preparedness personnel

6. What are the three public protective action strategies to protect the population from a hazardous material?

7. The public is more likely to comply with instructions to protect-in-place when certain factors are present, including these:

- Receipt of a timely and effective warning message
- Clear rationale for the decision to protect-in-place, as compared to an evacuation
- Credibility of emergency response personnel with the general public

What is another factor?

8. Two variables considered when evaluating structures for protection-in-place are the age of the building and the prevailing wind direction (the direction the wind blows the majority of the time for the time of year in your area). What is another important variable?

9. A limited-scale evacuation may be the best option for the IC under the following conditions:

- When the building is on fire
- When the hazmat is leaking inside the building and the material is flammable or toxic
- When the building occupants show signs or symptoms of acute illness and there is a known hazmat spill inside the structure

What is another condition?

10. Define the term "sick building syndrome" (SBS):

11. The following are all situations that may justify a full-scale evacuation except one. Which is the exception?

    a. Large leaks involving flammable and/or toxic gases from large-capacity storage containers and process units

    b. Large quantities of materials that could detonate or explode, damaging additional process units, structures, and storage containers in the immediate area

    c. Leaks and releases that are difficult to control and could increase in size or duration

    d. A situation in which the hazardous material has been totally released from its container and is dissipating

12. Four critical issues must be addressed and managed when the decision is made to commit to a full-scale public evacuation. Three issues are identified here:

- Alerting and notification
- Transportation
- Information (keeping displaced people informed)

What is the fourth critical issue?

Self-Evaluation

Review all your work in this lesson and note your stronger and weaker areas.

Overall I feel I did (very well / well / fair / not so well) on the acronyms and abbreviations exercise.

Overall I feel I did (very well / well / fair / not so well) on the self-test questions.

Overall I feel did (very well / well / fair / not so well) on the terminology exercise.

Overall I feel I did (very well / well / fair / not so well) on the summary and review questions.

When compared to the previous lessons, I think I performed (better / worse / equally well).

List two areas in this chapter in which you feel you could improve your skill or knowledge level:

1. _____

_____

_____

_____

2. _____

_____

_____

_____

Consider the following self-evaluation questions as they pertain to this chapter:

Am I taking effective notes?

Am I dedicating enough quality time to my studies?

Is anything distracting my focus?

Was any part of this chapter too advanced for me?

Did I find that I don't have enough background experience to sufficiently grasp certain subject areas?

For areas in which I did particularly well, was it because I'm particularly interested in that subject matter? How so?

Were some things easier to learn because I have prior experience in learning or working with the concepts or principles?

Did I find that certain portions of the textbook seem to be better organized and effective in explaining key points?

Identifying the Problem

Chapter Orientation

Open the textbook to the *Identifying the Problem* chapter. When you have finished looking through the chapter, respond to the following items. Use the textbook as you jot down your comments in the spaces provided below.

1. From your own experience, would you agree that a hazmat problem well defined is half-solved? Why or why not?

   _____

   _____

   _____

   _____

   _____

2. Reflect on your current level of knowledge or background experience about the topics covered in this chapter. Where have you read about, learned about, or applied this knowledge in the past?

   _____

   _____

   _____

   _____

   _____

3. Which sections or parts of this chapter strike you as looking especially interesting?

   _____

   _____

   _____

   _____

   _____

4. Which particular subjects in this chapter are important for a person in your position to master?

   _____

   _____

   _____

   _____

   _____

5. What do you predict will be the most difficult things for you to learn in this chapter?

   _____

   _____

   _____

   _____

   _____

Learning Objectives

Examine the *Identifying the Problem* chapter objectives and respond to the following question:

1. Which objectives in this chapter do you feel you can achieve right now, with a reasonable level of confidence?

_____

_____

_____

_____

As you read the sections of the chapter that deal with the objectives you have identified above, make sure your ideas and knowledge base match those of the authors. If they do not match, you should examine how your current understanding of the material differs from that of the authors. Depending on the level at which you wish to master the subject, discrepancies will have to be rectified and gaps will need to be filled.

Abbreviations and Acronyms

The following abbreviations and acronyms are used in the *Identifying the Problem* chapter:

AAR	Association of American Railroads
ANSI	American National Standards Institute
API	American Petroleum Institute
ASME	American Society of Mechanical Engineers
ASTM	American Society of Testing and Materials
CAS	Chemical Abstract Service
CGA	Compressed Gas Association
CHEMTREC®	Chemical Transportation Emergency Center
COFC	Container on Flat Car
CSI	Criticality Safety Index
CTC	Canadian Transport Commission
DOT	U.S. Department of Transportation
EPA	Environmental Protection Agency
ERG	*Emergency Response Guidebook*
GHS	Global Harmonization System
HMIG	*Hazardous Materials Identification Guide*
HMIS	Hazardous Materials Information System
HMT	Hazardous Materials Table

HRCQ	Highway Route Controlled Quantity
IBC	Intermediate Bulk Container
ICAO	International Civil Aviation Organization
IDLH	Immediately Dangerous to Life or Health
LC$_{50}$	Lethal Concentration, 50% Kill Rate
LEPC	Local Emergency Planning Committee
LOX	Liquid Oxygen
LPG	Liquefied Petroleum Gas
LSA	Low Specific Activity
MAWP	Maximum Allowable Working Pressure
NA	North America
NASFM	National Association of State Fire Marshals
NFPA	National Fire Protection Association
NOS	Not Otherwise Specified
NTSB	National Transportation Safety Board
ORM-D	Other Regulated Material
PCBs	Polychlorinated Biphenyls
PG	Packing Group
PHMSA	Pipeline and Hazardous Materials Safety Administration
PIH	Poison–Inhalation Hazard
RQ	Reportable Quantity
SARA	Superfund Amendments and Reauthorization Act
SCADA	Supervisory Control and Data Acquisition System
SCO	Surface-Contaminated Objects
SDS	Safety Data Sheet
SI	International System Units
STCC	Standard Transportation Commodity Code
TC	Transport Canada
TI	Transport Index
TIH	Toxic–Inhalation Hazard
TOFC	Trailer on Flat Car
ULD	Unit Load Devices
UN	United Nations
WMD	Weapons of Mass Destruction

Abbreviations and Acronyms Exercise

Fill in the blank using the abbreviations and acronyms from the preceding list. Use each abbreviation or acronym only once.

1. Product flows through many transmission pipeline systems are monitored through this computerized pipeline system: _____

2. Sometimes referred to as a chemical's "social security number," these sequentially assigned numbers identify specific chemicals and have no chemical significance: _____

3. SARA Title III requires facilities to notify this entity when on-site quantities exceed established threshold values:

_____

4. This federal agency requires leak detection, overfill protection, and cathode protection on underground storage tanks: _____

5. This common source of emergency response information helps first responders find emergency procedures quickly: _____

6. This classification for a chemical indicates the degree of danger associated with its transportation: _____

7. This federal agency investigated the 1971 Houston, TX, railroad derailment: _____

8. In 2013, OSHA updated the Hazard Communication Standard and adopted this to harmonize U.S. hazard communication regulations with those of the international community: _____

9. DOT regulations establish these as a class of low-level radioactive waste: _____ and _____

10. This can be an indicator for determining the external radiation hazard of an undamaged package and can be a starting point for determining whether damage has occurred: _____

11. This record identifies the hazard class or specifies that the material is forbidden in transportation and gives the proper shipping name or directs the user to the preferred proper shipping name: _____

12. This agency provides an online national pipeline mapping system that is an excellent tool for learning about certain types of pipelines in a given area: _____

13. These materials do not require placards: _____

14. This seven-digit number will be found on all shipping papers accompanying rail shipments of hazmats: _____

Study Session Overview

Use the textbook to help you answer the questions or work from memory.

1. Based on your current understanding of problem identification, explain why problem identification cannot be safely accomplished if responders have not first controlled the incident scene.

   _____

   _____

   _____

   _____

   _____

2. What is the primary lesson learned from the 1971 Houston, TX, railroad derailment and explosion?

   _____

   _____

   _____

   _____

   _____

3. Based on your current understanding of hazardous materials response, what is the rationale behind this statement: "If you can't identify, then try to classify?"

_____

_____

_____

_____

_____

4. Explain the relationship between distance and exposure risk when selecting an identification method (clues).

_____

_____

_____

_____

_____

Self-Test

Use the textbook to help you answer the questions or work from memory.

1. The identification process has three basic elements: recognition, identification, and:
   a. classification.
   b. verification.
   c. detection.
   d. description.

2. Which of the following occupancies and locations is a potential hazmat site?
   a. Water treatment plant
   b. Railroad tank car
   c. Residential apartment
   d. All of the above

3. Packaging used for transporting hazardous materials is regulated by:
   a. American Petroleum Institute (API).
   b. ASME, ASTM, and ANSI.
   c. Dept. of Transportation (DOT).
   d. NFPA 30.

4. The following are all examples of nonbulk packaging *except*:
   a. a ton container.
   b. a drum.
   c. a carboy.
   d. glass bottles inside a fiberboard box.

5. True / False: Bulk packages can be an integral part of a transport vehicle or packaging placed on or in a transport vehicle.

6. Which of the following is an example of a facility containment system?

    a. Storage bins or cabinets

    b. Dryers and degreasers

    c. Piping systems

    d. All of the above

7. Of the following materials, which is most likely to be found in a nonbulk bag?

    a. Radioactive materials

    b. Fertilizers

    c. Solvents

    d. Infectious disease samples

8. Which of the following would *not* be found in an aluminum drum?

    a. Materials that react with rust

    b. Combustible materials

    c. Caustic corrosives

    d. Pesticides

9. True / False: Almost any hazmat can be found inside a box.

10. Of the following materials, which is most likely to be found in a nonbulk carboy?

    a. Corrosive liquids

    b. Nonflammable pressurized gases

    c. Cryogenic liquids

    d. Explosives

11. Which of the following bulk packaging would contain liquefied gases, including chlorine?

    a. Super sacks

    b. Ton containers

    c. Polyethylene and steel tanks

    d. Portable bins

12. A DOT Spec 51 pressure tank container is most likely to contain:

    a. food-grade commodities.

    b. poison liquids.

    c. alcohols.

    d. pyrophoric liquids.

13. Service pressures ranging up to 2,400 psi and higher are characteristic of:

    a. tube modules.

    b. IMO Type 7 tanks.

    c. IM 101 tanks.

    d. None of the above.

14. A DOT-406 AL specification indicates that the cargo tank:

    a. is inspected annually.

    b. was manufactured prior to 1990.

    c. is constructed of aluminum.

    d. is registered in Alabama.

Match each cargo tank truck in Column A to its specification type in Column B.

Column A	Column B
Cargo Tank Truck	**Contents/Hazard Class**
_____ 15. Atmospheric pressure cargo tank truck	**A.** MC-330/MC-331
_____ 16. Low-pressure chemical cargo tank truck	**B.** MC-306/DOT-406
_____ 17. Corrosive cargo tank truck	**C.** MC-338
_____ 18. High-pressure cargo tank truck	**D.** MC-312/DOT-412
_____ 19. Cryogenic liquid cargo tank truck	**E.** MC-307/DOT-407

20. Heated material cargo tank trucks are required to be placarded as:

a. "HOT" material.

b. "Dangerous" material.

c. "Combustible" material.

d. "Molten Liquid" material.

21. Reporting marks and number (e.g., GATX 12345) can be used to obtain information about:

a. the contents of the car.

b. the tank car's test pressure.

c. the type of material used in tank construction.

d. the builder's name, tank specification, and class designation.

Match each railroad tank car in Column A to the materials it carries, listed in Column B.

Column A	Column B
Railroad Tank Car	**Contents/Hazard Class**
_____ 22. Nonpressure tank car	**A.** Liquid oxygen, liquid hydrogen, liquid argon
_____ 23. Pressure tank car	**B.** Liquid poisons, organic peroxides, vegetable oils
_____ 24. Cryogenic liquid tank car	**C.** Chlorine, anhydrous ammonia, LPG
_____ 25. Pneumatically unloaded covered hopper car	**D.** Pellets, caustic flake, resin powder, flour

26. Most low-level radioactive waste, such as contaminated protective clothing, is shipped in:

a. excepted packaging.

b. industrial packaging.

c. Type A packaging.

d. Type B packaging.

27. Potentially life-endangering amounts of radioactive material, such as spent nuclear fuel, that could pose a significant risk if released during an accident are transported in:

a. excepted packaging.

b. industrial packaging.

c. Type A packaging.

d. Type B packaging.

28. The most common storage tank found in the petroleum industry is the:

    a. cone roof tank.

    b. covered floating roof tank.

    c. open floating roof tank.

    d. open floating roof tank with geodesic dome.

29. At chemical facilities, underground storage tanks may store:

    a. only combustible liquids.

    b. only liquefied petroleum gases.

    c. flammable and nonflammable liquefied gases.

    d. virtually any hazardous or nonhazardous liquid.

30. The signal words found on agricultural chemicals and pesticide container labels indicate:

    a. level of toxicity.

    b. concentration.

    c. ratio of active to inert ingredients.

    d. hazard class.

31. Which of the following is the only reliable way to identify cylinder contents?

    a. Color codes

    b. CAS number

    c. Product information stencil

    d. DOT label attached to the cylinder head

32. Pipeline markers must identify the pipeline contents, the pipeline operator, and:

    a. the four-digit identification number.

    b. an emergency telephone number.

    c. the CAS number.

    d. the DOT hazard class.

33. The yellow quadrant in the NFPA 704 marking system indicates:

    a. PCBs.

    b. reactivity.

    c. toxicity.

    d. radioactivity.

34. The hazmat placarding and labeling requirements within the U.S. are regulated by the:

    a. EPA.

    b. API.

    c. DOT.

    d. OSHA.

35. Placards that are applied to freight containers, cargo tanks, and portable tank containers are approximately _____ in size.

   a. 4 inches square

   b. 11 inches square

   c. 18 inches square

   d. 24 inches square

36. A package with a radioactive White-I label will (normally) contain:

   a. fissile Class III materials.

   b. materials rated at a maximum allowable TI = 1.

   c. materials with high radiation levels.

   d. materials with extremely low or almost no levels of radiation.

37. What does the Packing Group entry on shipping papers indicate?

   a. Type of packaging

   b. Degree of danger

   c. Hazard class

   d. Hazardous waste classification

38. Shipping papers must include which type of emergency response information?

   a. Personnel protective measures

   b. First aid measures

   c. Emergency actions involving fire

   d. All of the above

Practice

1. Obtain a copy of the latest DOT *Emergency Response Guidebook* (ERG; print or online version). Practice using the ERG to identify hazardous materials by identification number.

DOT Number	Chemical Name
1051	
1256	
1114	
1075	
2055	
2810	
1198	
2790	
1050	
1972	
1760	
1079	
1203	
1282	
1005	

2. Locate an online CAS Registry Number Search service (e.g., one such service is offered by NIST; http://webbook.nist.gov/chemistry/cas-ser.html). Practice using the CAS number to identify hazardous materials.

CAS Number	Chemical Name
506-77-4	
8030-30-6	
8014-95-7	
7664-38-2	
1310-73-2	

Study Group Activity

Working with a partner, take a few field trips to various industries within your community, as well as to railroad yards and pipeline crossover locations. Identify the marking systems used and practice locating and interpreting the information provided by the markings. Record your observations. Which ones are the easiest to locate and interpret? Which ones are the most difficult or provide the least useful information for emergency responders? Discuss your experiences with your study group.

Study Group Learning Through Inquiry Question 6–1

You are the fire marshal within your community. The Local Emergency Planning Committee (LEPC) has proposed an ordinance to the local elected officials that will mandate a facility marking system be implemented for all SARA Title III reporting facilities within the community. The primary objective of this system would be to allow emergency responders to (1) identify those facilities that manufacture, store, handle, and use hazardous materials within the community and (2) allow emergency responders to become more familiar with such facilities.

Although the fire chief favors the proposal because of fire fighter health and safety concerns, the police chief has raised some substantive security issues with respect to identifying such facilities to the public. The police chief points out that hazardous materials stored at legitimate and legal facilities can be stolen and used to make an explosive device or as feedstock to make illegal drugs. Such facilities could also be accessed during a riot. Why make it easier for the "bad guys" by telling them where the hazmat is stored?

The fire chief has turned the LEPC's proposal over to you and is now awaiting your recommendations and suggestions as to how the fire department should pursue the issue. He makes it clear that he wants a marking system that will improve fire fighter safety. He also makes it clear that he has worked hard to develop a good working relationship with the police chief and he wants to be responsive to the police chief's concerns. Any recommendation that you develop must address the security concerns.

Using the limited background information provided and the information discussed in the *Identifying the Problem* chapter, answer the following questions:

1. Which "off the shelf," nationally recognized systems for identifying locations associated with hazardous materials are available that could be adopted? What would be the advantages and disadvantages of adopting one system over another?

2. How can the facility marking system program be implemented from a safety perspective while minimizing the security concerns expressed by the police department?

3. Assuming that you believe the safety benefits to fire fighters of implementing a marking system far outweigh the negative aspects from a security point of view, how would you make your case to the police chief?

Study Group Learning Through Inquiry Question 6–2

You are the incident commander at the scene of a tractor-trailer incident involving the leaking of an unknown liquid substance. It is approximately 4:00 P.M. on a cold February weekday. The incident was reported by the truck driver, who noticed liquid leaking from his truck while looking at his rear view mirror and driving down the interstate highway. The driver pulled off the interstate and parked his vehicle on a side street in a heavily congested area and dialed 911 for assistance.

Your HMRT has isolated the immediate area and interviewed the driver, who is very cooperative but has limited knowledge of his cargo. The manifest shows that the truck contains fifty-seven 55-gallon drums of mixed hazardous waste. Most of the drums contain different pesticide solutions.

After evaluating the hazards and risk, you decide to send a recon team to the truck to inspect the vehicle and recover a sample. As a precaution, you have decided to evacuate several nearby stores. The owners are very unhappy about this because 5:00 P.M. to 7:00 P.M. is their best sales period.

While this operation is being conducted, several TV news vehicles show up and meet with your PIO. They indicate that they will be going live for the 6 o'clock news broadcast. The time is about 5:30 P.M. You agree to have your PIO conduct a preliminary press briefing; however, you inform the PIO that recent hazmat incidents along the interstate highway corridor have generated many citizen complaints concerning the need to close down the highway. There is a great deal of local sensitivity to this issue, and he should anticipate that questions might be raised during the interviews.

After conducting the first press briefing, your recon team reports that the leaking liquid appears to be water, which is melting from ice trapped on the top of the hazardous waste drums. A sample tested on the scene using a HazCat® Kit confirms that the leaking liquid is, in fact, uncontaminated water.

Based on the background information provided and the information in the *Identifying the Problem* chapter, answer the following questions:

1. Which methods of identification should be available to the HMRT at this incident that would help confirm that the cargo is hazardous waste? How could the HMRT identify the constituents of the individual hazardous waste drums?

2. How would you brief the media on this issue if you conducted the interview *before* you learned that the material leaking was water? What are the three key points you would want to make in your briefing?

3. Assuming that you already briefed the media once (question 2), how would you brief the media *after* you discovered that the leaking material was water? Would you specifically tell the media that the material was water and not hazardous waste?

4. How do you think this incident would look to the general public if the evacuation issue was raised by a reporter on live television? You justified the need to evacuate and block roads in the first interview, but now it turns out that you took this action for leaking water! Isn't this just another case of the fire department overreacting to the problem?

Summary and Review

1. Where do most hazardous materials releases occur, in facilities or during transportation?

2. List at least three basic recognition clues (e.g., occupancy and location):

3. Identify those clues that can be used to identify possible weapons of mass destruction:

4. Of the various identification methods, which one poses the highest risk to responders? Why?

5. Which federal organization regulates packaging used for transporting hazardous materials?

6. Packing is divided into three general groups:
   1. Nonbulk packaging
   2. _____
   3. _____

7. What are overpack drums used for?

8. In which facilities are carboys typically found, and what can be found inside them?

9. In which type of cylinder are cryogenic liquids found?

10. On a transport vehicle, you locate a cylindrical pressure tank approximately 3 feet in diameter and 8 feet long with concave heads. Which type of container is this, and what is its likely contents/hazard class?

11. Unless under a DOT exemption, flexible containers or "super sacks" are *not* authorized to contain:
    a. explosives.
    b. oxidizers and organic peroxides.
    c. liquid hazardous materials.
    d. water treatment chemicals.

12. You locate a tank container consisting of seamless steel cylinders from 9 to 48 inches in diameter, permanently mounted inside an open frame. A box-like compartment at one end encloses all the valves. What are this container's likely contents?

13. What does the capacity stencil on a railroad tank car tell you?

14. Which of the following radioactive material packaging is *not* designed to withstand the forces of an accident?
    a. Excepted packaging
    b. Industrial packaging
    c. Type A packaging
    d. Type B packaging

15. There are three ways to positively identify a pesticide. One way is by product name. What are the other two ways?

16. What is the purpose of four-digit identification numbers?

17. Placards and labels provide recognition and general hazard classification by way of four indicators:
    1. Colored background

    2.

    3.

    4.

18. Which of the following types of liquid storage tanks may contain solvents, oxidizers, and corrosives?
    a. Cone roof tank
    b. Horizontal storage tank
    c. Covered floating roof tank
    d. Both a and b

19. If the airbill is not attached to the outside of packages being transported by air, where should you look for it?

20. True / False: All shipping papers must contain an emergency response telephone number.

Self-Evaluation

Review all your work in this lesson and note your stronger and weaker areas.

Overall I feel I did (very well / well / fair / not so well) on the acronyms and abbreviations exercise.

Overall I feel I did (very well / well / fair / not so well) on the self-test questions.

Overall I feel I did (very well / well / fair / not so well) on the summary and review questions.

When compared to the previous lessons, I think I performed (better / worse / equally well).

List two areas in this chapter in which you feel you could improve your skill or knowledge level:

1. _____

   _____

   _____

   _____

2. _____

_____

_____

_____

Consider the following self-evaluation questions as they pertain to this chapter:

Am I taking effective notes?

Am I dedicating enough quality time to my studies?

Is anything distracting my focus?

Was any part of this chapter too advanced for me?

Did I find that I don't have enough background experience to sufficiently grasp certain subject areas?

For areas in which I did particularly well, was it because I'm particularly interested in that subject matter? How so?

Were some things easier to learn because I have prior experience in learning or working with the concepts or principles?

Did I find that certain portions of the textbook seem to be better organized and effective in explaining key points?

Hazard Assessment and Risk Evaluation

Chapter Orientation

Open the textbook to the *Hazard Assessment and Risk Evaluation* chapter. When you have finished looking through the chapter, respond to the following items. Use the textbook as you jot down your comments in the spaces provided below.

1. Stated as simply as possible, what is the objective of hazard and risk evaluation?

   _____

   _____

   _____

   _____

   _____

2. Reflect on your current level of knowledge or background experience about the topics covered in this chapter. Where have you read about, learned about, or applied this knowledge in the past?

   _____

   _____

   _____

   _____

   _____

3. Which sections or parts of this chapter strike you as looking especially interesting?

   _____

   _____

   _____

   _____

   _____

4. Which particular subjects in this chapter are important for a person in your position to master?

   _____

   _____

   _____

   _____

   _____

5. What do you predict will be the most difficult things for you to learn in this chapter?

   _____

   _____

   _____

   _____

   _____

Learning Objectives

Examine the *Hazard Assessment and Risk Evaluation* chapter objectives and respond to the following question:

1. Which objectives in this chapter do you feel you can achieve right now, with a reasonable level of confidence?

_____

_____

_____

_____

_____

As you read the sections of the chapter that deal with the objectives you have identified above, make sure your ideas and knowledge base match those of the authors. If they do not match, you should examine how your current understanding of the material differs from that of the authors. Depending on the level at which you wish to master the subject, discrepancies will have to be rectified and gaps will need to be filled.

Abbreviations and Acronyms

The following abbreviations and acronyms are used in the *Hazard Assessment and Risk Evaluation* chapter:

ACC	American Chemistry Council
AHJ	Authority Having Jurisdiction
ALOHA	Aerial Locations of Hazardous Atmospheres
ASPCA	American Society for the Prevention of Cruelty to Animals
ATSDR	Agency for Toxic Substances and Disease Registry
CAMEO®	Computer-Assisted Management of Emergency Operations
CANUTEC	Canadian Transport Emergency Centre
CDC	Centers for Disease Control and Prevention
CEPPO	Chemical Emergency Prevention and Preparedness Office (EPA)
CGI	Combustible Gas Indicator
CHARM®	Complex Hazardous Air Release Model (software)
CHEMTREC®	Chemical Transportation Emergency Center
CHRIS	Chemical Hazards Response Information
CMS	Chip Measurement System
COMSEC	Communications Security
CPM	Counts per Minute
eV	Electron-Volts

FID	Flame Ionization Detector
FT-IR	Fourier Transform Infrared Spectrometry
GC	Gas Chromatograph
GHBMO	General Hazardous Materials Emergency Behavior Model
GHS	Globally Harmonized System of Classification and Labeling of Chemicals
HDPE	High-Density Polyethylene
HHA	Hand-Held Immunoassay
IMS	Ion Mobility Spectroscopy
IP	Ionization Potential
JHAT	Joint Hazard Assessment Team
JTTF	Joint Terrorism Task Force
LDL	Lower Detection Limit
LEL	Lower Explosive Level
LEPC	Local Emergency Planning Committee
MARPLOT	Mapping Applications for Response, Planning, and Local Operational Tasks
MS	Mass Spectrometer
MSST	Maximum Safe Storage Temperature
NEC	National Electrical Code
NPIC	National Pesticide Information Center
NRC	National Response Center
NTSB	National Transportation Safety Board
PCR	Polymerase Chain Reaction
PDA	Personal Data Assistants
pH	Power of Hydrogen
PID	Photo-ionization Detector
RQ	Reportable Quantity
SADT	Self-Accelerating Decomposition Temperature
SETIQ	Emergency Transportation System for the Chemical Industry
USAMRID	U.S. Army Medical Research Institute of Infectious Diseases
USSS	U.S. Secret Service
UV	Ultraviolet
VOC	Volatile Organic Compound
WISER	Wireless Information System for Emergency Responders

Abbreviations and Acronyms Exercise

For each of the following sentences, write in the correct abbreviation or acronym (from the preceding list) so that that sentence makes sense. Use each abbreviation or acronym only once.

1. There are two primary ionizing detectors used in the field: the flame ionization detector and the

_____.

2. Flame ionization detectors operate in two modes: the survey mode and the _____ mode.

3. The _____ is usually coupled with a gas chromatograph and is the identifying portion of the device.

4. _____ technology allows for the specific identification of liquid and solid samples.

5. Radiation survey instruments are read in units of _____ .

6. The _____ scale ranges from 0 to 14.

7. Irreversible decomposition of an organic peroxide begins when the _____ is reached.

8. The _____ is used to plot releases of hazardous materials vapors.

9. When using a photo-ionization detector, the sample is exposed to a(n) _____ lamp, which ionizes the sample.

Match each agency abbreviation in Column A to its description in Column B. Use the textbook to help you or work from memory.

Column A	Column B
Agency	**Description**
_____ **10.** SETIQ	**A.** The leading federal public health agency for hazmat incidents.
_____ **11.** ACC	**B.** Provides assistance in identification and establishing contact with shippers and manufacturers of hazardous materials that originate in Canada.
_____ **12.** NRC	**C.** Operates the Animal Poison Control Center and provides consultation in the diagnosis and treatment of suspected or actual animal poisonings or chemical contamination.
_____ **13.** ATSDR	**D.** A service of the Mexico National Association of Chemical Industries.
_____ **14.** CANUTEC	**E.** Federal agency that investigates transportation accidents and promulgates safety recommendations.
_____ **15.** CHEMTREC®	**F.** Operates CHEMTREC®, the most recognized emergency information center in the United States.
_____ **16.** CEPPO	**G.** Provides information on pesticide-related health/toxicity questions, properties, and minor clean-up to physicians, veterinarians, responders, and the general public.
_____ **17.** ASPCA	**H.** Provides a number of emergency and nonemergency services, including emergency response information and emergency communications.
_____ **18.** NPIC	**I.** The federal government's central reporting point for all oil, chemical, radiological, biological, and etiological releases into the environment within the United States and its territories.
_____ **19.** NTSB	**J.** An organization within the EPA that provides a good starting point for gathering hazard information.

Study Session Overview

Use the textbook to help you answer the questions or work from memory.

1. Based on what you have read in this chapter and have previously learned about hazardous materials incident management, how might the outcome of the 1989 hazmat train derailment (described in this chapter) have been different had the IC played a more active role during the hazard and risk assessment process?

_____

_____

_____

_____

_____

2. In your own words, define what a hazard is and provide several examples of hazards.

_____

_____

_____

_____

_____

3. Define risk. Why can't we evaluate risk based solely on knowledge and understanding of the hazards involved?

_____

_____

_____

_____

_____

4. Which of the hazard data and information sources identified in this chapter do you have the most experience with? Which ones would you like to become more familiar with?

_____

_____

_____

_____

_____

5. The "Rule of Threes" means using several types of detection technologies to classify or identify the hazard. What is the rationale for this practice?

_____

_____

_____

_____

_____

6. In your own words, explain why defensive tactics are always preferable over offensive tactics if they can accomplish the same objectives in a timely manner.

_____

_____

_____

_____

_____

7. What is the nonintervention mode, and when is it ideally implemented?

_____

_____

_____

_____

Self-Test

Use the textbook to help you answer the questions or work from memory.

1. What is the temperature of a material within its container called?
   a. Ambient temperature
   b. Normal physical state
   c. Temperature of product
   d. Critical temperature

2. Which of the following properties is the most significant for evaluating spill control options and clean-up procedures for waterborne releases?
   a. Specific gravity
   b. Boiling point
   c. Vapor pressure
   d. Concentration

3. Critical temperature and critical pressure are both terms that relate to:
   a. the distance that gases and vapors will travel.
   b. the process of liquefying gases.
   c. the evaluation of flammability hazards of a material.
   d. the point at which the vapor pressure of a liquid equals atmospheric pressure.

4. Miscibility is the ability of materials to:
   a. float on water.
   b. absorb into a permeable solid.
   c. dissolve into a uniform mixture.
   d. change chemically into another substance.

5. Which of the following is most significant in determining the temperature at which the vapors from a flammable liquid are readily available and may ignite?
   a. Flash point
   b. Fire point
   c. Maximum safe storage temperature (MSST)
   d. Self-accelerating decomposition temperature (SADT)

6. What is the percentage of an acid or base dissolved in water called?
   a. Strength
   b. pH
   c. Water reactivity
   d. Concentration

7. Which of the following is another term for *caustic*?
   a. Air reactive
   b. Corrosive
   c. Acid
   d. Base

8. The time it takes for the activity of a radioactive material to decrease to one half of its initial value through radioactive decay is called the:

    a. dose.

    b. dose rate.

    c. half-dose.

    d. half-life.

9. Chemical warfare agents such as tabun, sarin, soman, and VX are all examples of:

    a. nerve agents.

    b. choking agents.

    c. vesicants (blister agents).

    d. pathogens.

10. Which of the following terms refers to the length of time a chemical agent remains as a liquid?

    a. Strength

    b. Activity

    c. Persistence

    d. Oxidation ability

11. The guidebooks and manuals listed in this chapter are written for which target audience?

    a. First Responders—Awareness and Operations levels

    b. Hazardous Materials Technicians and Specialists

    c. Incident Commanders

    d. Product Specialists

12. Which of the following agencies must be notified by the responsible party (i.e., the spiller) if a hazardous materials release exceeds the reportable quantity (RQ) provisions of CERCLA?

    a. EPA Chemical Emergency Preparedness and Prevention Office (CEPPO)

    b. Department of Transportation National Response Center (NRC)

    c. Federal Emergency Management Agency (FEMA)

    d. U.S. Chemical Safety and Hazard Investigation Board

13. There are *no* regulatory requirements that safety data sheets (SDS) provide:

    a. fire and explosion data.

    b. hazard ingredient statement.

    c. spill and leak control procedures.

    d. standard language and terminology.

14. The inherent safety of a direct-reading instrument pertains to the ability of the device to:

    a. operate in hazardous atmospheres.

    b. select slight changes in product concentrations.

    c. monitor for both very low and very high concentrations.

    d. determine the exact contaminant present.

15. Which of the following devices monitors the accumulated radiation dose received by an individual?

    a. Radiation pager

    b. Radiation meter

    c. Dosimeter

    d. Passive sensor

16. Which of the following devices must be calibrated prior to use to compensate for altitude and barometric pressure?

    a. pH meter

    b. Oxygen monitor

    c. Ion chamber

    d. Colorimetric indicator tubes

17. Response curves are required to read:

    a. combustible gas indicators (CGIs).

    b. flame ionization detectors.

    c. toxic gas sensors.

    d. Fourier-transform infrared spectrometry (FT-IR).

18. Which of the following devices can be susceptible to false readings if hand-carried or moved around?

    a. Chemical test strip

    b. Mercury detector

    c. Photo-ionization detector (PID)

    d. Geiger-Müeller tube

19. Which of the following monitoring instruments is used to monitor for explosives?

    a. Fourier-transform infrared spectrometry (FT-IR)

    b. Photo-ionization detector

    c. Colorimetric indicator tube

    d. Combustible gas indicator

20. Which of the following is given monitoring priority if there is any doubt that the hazard is present?

    a. Radiation

    b. Flammability

    c. Oxygen-deficient/enriched atmosphere

    d. Toxicity

21. If the incident involves a confined-space scenario, OSHA clearly outlines the first monitoring priority as:

    a. radiation.

    b. flammability.

    c. oxygen-deficient/enriched atmosphere.

    d. toxicity.

22. The following guidelines for collecting evidence samples are all true except one. What is the exception?

    a. Sampling tools and gloves must be used only one time for each sample.

    b. Samples collected for product identification may be used for evidentiary purposes.

    c. Control blanks should be provided as part of the sampling process.

    d. Sample containers that are certified as "clean" will have a letter stating that they are cleaned to some specification.

23. What is the last stage in the Hazardous Materials Emergency Model?

    a. Breached

    b. Engulfed

    c. Over-stressed

    d. Stabilized

24. The General Hazardous Materials Behavior Model (GHBMO) includes six events: stress, breach, release, engulf, impinge, and:

    a. harm.

    b. spill.

    c. force.

    d. escalate.

25. True / False: Chemical stress, mechanical stress, and thermal stress can occur in combination with each other.

26. Which of the following types of breach behaviors is commonly associated with catastrophic BLEVE scenarios?

    a. Disintegration

    b. Failure of container attachments

    c. Runaway cracking

    d. Container punctures

27. Which of the following types of releases usually offers responders adequate time to develop prolonged countermeasures?

    a. Detonation

    b. Violent rupture

    c. Rapid relief

    d. Spills or leaks

28. Impingements are categorized based on:

    a. duration.

    b. dispersion patterns.

    c. rate of release.

    d. the type of container.

29. Etiologic harm results from:

    a. exposure to poisons.

    b. exposure to simple asphyxiants.

    c. exposure to corrosive materials.

    d. exposure to biological materials.

30. The relative ability of a metal to bend or stretch without cracking is called:

    a. deformation.

    b. ductility.

    c. density.

    d. deflection.

31. True / False: Tank car dent gauges cannot be used for assessing dents on cargo tank trucks due to differences in shell metal and thickness.

32. True / False: The potential for ignition within a sewer collection system will be greatest at points where flammable liquids may enter or where entry is possible.

Practice

1. Visit the CHEMTREC® web site (http://www.chemtrec.com). To become more familiar with its services, use the web site to answer following questions.

   a. What does the acronym CHEMTREC® stand for?

   _____

   _____

   _____

   _____

   b. What are some of the 24/7 information resources that CHEMTREC® has to assist callers with an incident?

   _____

   _____

   _____

   _____

   c. Which types of information will the CHEMTREC® emergency service specialists request when you call them for assistance?

   _____

   _____

   _____

   _____

   d. Will CHEMTREC® notify other federal, state, or local authorities for you in case of a hazmat spill?

   _____

   _____

   _____

   _____

   e. Does CHEMTREC® assist with handling a medical exposure?

   _____

   _____

   _____

   _____

2. Assume that you have been tasked with developing a list of product and container specialists for terrorism and WMD agents. How will you proceed?

   _____

   _____

   _____

   _____

3. Based on the information in this chapter and using outside resources, develop a proposal to equip your department with a state-of-the-art sampling equipment kit. You do not need to research costs, but you should be able to substantiate the need for the equipment or services you are proposing to acquire.

_____

_____

_____

_____

4. Using the DOT *Emergency Response Guidebook* (ERG; print or online version) or other resources, estimate the area potentially impacted by a release of the following hazardous materials. Assume you have a large release, such as a one-ton cylinder, a tank truck, or a railcar.

Sulfur mustard	Fire isolation:	
	Spill isolation: (meters)	Downwind: (miles)
Phosgene	Fire isolation:	
	Isolation: (meters)	Downwind: (miles)
Sarin	Fire isolation:	
	Isolation: (meters)	Downwind: (miles)
Chlorine	Fire isolation:	
	Isolation: (meters)	Downwind: (miles)

Important Terminology

Practice your recall of the important terms in this chapter by answering the following questions.

Select the correct answer.

1. Which materials contain carbon atoms?
   a. Organic materials
   b. Inorganic materials

2. Which hydrocarbons contain a benzene "ring?"
   a. Aromatic hydrocarbons
   b. Halogenated hydrocarbons

3. Of these two hydrocarbons, which is more chemically active and considered more hazardous?
   a. Saturated hydrocarbons
   b. Unsaturated hydrocarbons

4. Neutralization, decomposition, and oxidation are all examples of:
   a. chemical change.
   b. chemical interaction.

5. What is the minimum temperature at which a liquid gives off sufficient vapors to ignite and sustain combustion?

    a. Flash point

    b. Fire point

6. Pyrophoric materials are:

    a. air reactive.

    b. chemical reactive.

    c. water reactive.

7. A(n) _____ is used to control the rate of a chemical reaction by either speeding it up or slowing it down.

    a. catalyst

    b. inhibitor

8. Which compounds have a pH < 7?

    a. Acids

    b. Bases

On the line next to each physical property in Column A, print the letter of its definition from Column B.

Column A	Column B
Physical Property	**Definitions**
_____ **9.** Vapor density	**A.** The temperature at which a liquid changes its phase to a vapor or gas.
_____ **10.** Boiling point	**B.** The weight of a pure vapor or gas compared with the weight of an equal volume of dry air at the same temperature and pressure.
_____ **11.** Melting point	**C.** The weight of a solid or liquid material as compared with the weight of an equal volume of water.
_____ **12.** Sublimation	**D.** The ease with which a liquid or solid can pass into the vapor state.
_____ **13.** Volatility	**E.** The ability of a substance to change from the solid to the vapor phase without passing through the liquid phase.
_____ **14.** Viscosity	**F.** The temperature at which a solid changes its phase to a liquid. (This temperature may also be the freezing point depending on the direction of the change.)
_____ **15.** Specific gravity	**G.** Measurement of the thickness of a liquid and its ability to flow.

Match each agent in Column A to the correct examples in Column B.

Column A	Column B
Agent	**Examples of the Agent**
_____ **16.** Nerve agents	**A.** Phosgene, chlorine
_____ **17.** Choking agents	**B.** Mustard, lewisite
_____ **18.** Blood agents	**C.** Tabun, sarin
_____ **19.** Blister agents (Vesicants)	**D.** Hydrogen cyanide, cyanogens, chloride

Unscramble Puzzle

Use these hints to unscramble each term.

20. Xretuim _____
Substance made up of two or more elements or compounds, physically mixed together.

21. Rusylr_____
Pourable mixture of a solid and a liquid.

22. Oyrandobsch _____
Compounds primarily made up of hydrogen and carbon.

23. Eneetlm _____
Pure substance that cannot be broken down into simpler substances by chemical means.

24. Numoopcd _____
Chemical combination of two or more elements, either the same elements or different ones, that is electrically neutral.

25. Tunoliso _____
Mixture in which all of the ingredients are completely dissolved.

Study Group Activity

1. In the *Hazard Assessment and Risk Evaluation* chapter, the authors note that responders may find themselves overwhelmed by large amounts of data and information available from a multitude of sources, or that they may rely on only one digital-based reference source without verifying the accuracy of that information. As a group, discuss and exchange ideas and solutions for managing and prioritizing information needs.

2. Ludwig Benner is a pioneer in the field of hazardous materials. He developed both the DECIDE process for analyzing an incident and the General Hazardous Materials Behavior Model (GEBMO) for hazardous materials. Benner has recorded in an essay his personal experiences investigating hazmat emergencies that led to the development of both these models. This "Story of GEBMO" is archived with other personal papers at http://www.bjr05.net/models/GEBMO.html (or search the terms "Hazardous Materials Behavior Model Benner"). Individually, group members should read this essay carefully and then discuss their thoughts on the paper as a group.

3. Assume you are called to an incident involving a release of gasoline into a sewer collection system. Half of the group should prepare a presentation on how to determine whether the source of the problem is a spill or dumping of the gasoline directly into the sewer collection system; the other half of the group should prepare a presentation on how to determine whether the source of the problem is a subsurface release, such as a breached underground storage tank or pipeline. Both groups should then make and discuss their presentations.

Study Group Learning Through Inquiry Scenario 7–1

You are the Incident Commander on the scene of an overturned 7,000-gallon MC-307/DOT-407 cargo tank truck carrying toluene. The vehicle is located at an interchange of two major highways. The tank is leaking at a large rate from the manway area, and a large volume of product has pooled in and around the vehicle.

Based on this limited information, answer the following questions using the material from the *Hazard Assessment and Risk Evaluation* chapter.

1. Which hazards are present?

2. Which types of monitoring instruments should be used? Be specific. Why would you use the instrument that you selected rather than another type that is available? Which other air monitoring instruments could you use to confirm the presence and concentrations of toluene?

3. Nearby businesses are complaining about strong odors in the area and in their buildings. The TLV/TWA for toluene is 100 ppm and the IDLH value is 2,000 ppm. Using the combustible gas indicator available within your organization, first responders are getting readings of 5% of the lower explosive limit (LEL) outside of these structures. The CGI is calibrated on pentane and the conversion factor is 1.2, with a 25% margin of error. Is there a hazard to unprotected individuals within the area where the reading was obtained? How would you proceed?

4. Which risks are involved in this incident? How would you justify taking or not taking the risks you described based on the hazards present? How would you describe these risks to the businesses whose owners are complaining about the strong odor of toluene?

Study Group Learning Through Inquiry Scenario 7–2

You are the Incident Commander and arrive on the scene of a confined-space incident involving a man trapped inside a vertical storage tank that is 27 feet high and approximately 18 feet in diameter. The only manway on the tank is 36 inches in diameter and is on the top of the tank. The facility manager informs you that the tank was being repaired by an outside contractor. The tank is empty and was cleaned prior to entry, but previously contained ethylene dibromide (EDB). You send a fully protected HMRT member to the top of the tank to look inside. She reports back to you that the victim is motionless, in a seated position, and has an SCBA unit on with the face piece in place. You confirm that the contractor has been inside the tank approximately 10 minutes.

Using the background information provided and the information discussed in the *Hazard Assessment and Risk Evaluation* chapter, answer the following questions.

1. Would you attempt a confined-space rescue based on the information you have been provided? If so, which precautions would you take based on what you have learned in this chapter? (Also review pages in the *Health and Safety* chapter.)

2. Would you attempt a rescue if the contractor had been in the tank for 15 minutes? If you answered yes, on what do you base this decision? If you answered no, would you attempt a rescue if the person trapped inside the tank was an off-duty fire fighter and you knew him personally? What if the person were a member of your family—would this influence your decision?

3. Would you attempt a rescue if the contractor had been in the tank for 25 minutes? What if a member of your emergency response team personally knew the contractor—would this change your answer?

Summary and Review

1. Risk levels are variable and change from incident to incident. Factors that influence the level of risk include the hazardous nature of the material(s) involved and the quantity of the material. Identify at least two other variables that influence the level of risk:

2. Given the fact that anhydrous ammonia has a molecular weight of 17, how do we know that it will rise if released?

3. Given the evaporation rates of the following materials, which one will evaporate the most slowly?
  a. Mineral spirits = 0.1
  b. Xylene = 0.6
  c. Methyl ethyl ketone = 3.8
  d. Acetone = 5.6

4. Which of the following statements about vapor pressure is true?
  a. The vapor pressure of a substance at 100°F is always higher than the vapor pressure at 68°F.
  b. Vapor pressures reported in millimeters of mercury (mm Hg) are usually very low pressures.
  c. The lower the boiling point of a liquid, the greater vapor pressure at a given temperature.
  d. All of the above

5. What is the relationship between expansion ratio and the amount of gas produced by evaporation?

6. Which of the following is a significant property in evaluating the selection of control and extinguishing agents, including the use of water and Class B firefighting foams?
  a. Relative gas density
  b. Boiling point
  c. Solubility
  d. Viscosity

7. What happens to the flammable range if a gas or vapor is released into an oxygen-enriched atmosphere?

8. Toxic byproducts of the combustion process are based on the burning material(s). If nitrogen is present, incomplete combustion can produce:

9. Reactive materials include materials that decompose spontaneously, polymerize, or otherwise self-react, such as oxidizers. Identify at least one other reactive material:

10. Assume you are the incident commander at a hazmat release. You need to initiate a conference phone call between your on-scene responders and company representatives. You also need to have SDSs faxed to you. Which organization can do this for you?

11. Identify at least two criteria that should be used when evaluating electronic-based information sources:

12. What is "instrument response time?"

13. If air monitoring provides no information on the identity or hazard class of the unknown material, what should the next step be?

14. Which method can be used to help responders avoid becoming overwhelmed by the volume of hazard information collected during the course of an incident?

15. Examine the Hazardous Materials Emergency Model on in this chapter. At which stages are emergency response countermeasures applied to influence the progression of events?

16. The following chart is based on the General Hazardous Materials Behavior Model. Some of the event categories have been filled in. Fill in at least one additional blank for each event:

Event					
Stress	**Breach**	**Release**	**Engulf**	**Impinge**	**Harm**
Thermal	Disintegration	Detonation	Cloud	Short term	Thermal
Radiation	Punctures	Leak	Cone	Medium term	Etiologic

17. Assume you are at an incident involving a pressurized bulk transportation container. What should you do if you are unsure of the container or how the container is likely to breach?

18. The movement of hazardous materials through soil depends on the viscosity of the liquid. Identify two other variables that influence this movement:

19. Although combustible gas indicators (CGIs) are excellent tools for evaluating flammable atmospheres, they may not be very effective for assessing low-level flammable concentrations, such as those found with subsurface and sewer spills. Why? Which instruments are better suited for these scenarios?

20. The probability of an explosion within an underground space will depend on two factors: 1) that a flammable atmosphere exists, and 2):

21. List at least five site safety procedures for hydrocarbon spills into sewer collection systems.

Self-Evaluation

Review all your work in this lesson and note your stronger and weaker areas.

Overall I feel I did (very well / well / fair / not so well) on the acronyms and abbreviations exercise.

Overall I feel I did (very well / well / fair / not so well) on the self-test questions.

Overall I feel did (very well / well / fair / not so well) on the terminology exercise.

Overall I feel I did (very well / well / fair / not so well) on the summary and review questions.

When compared to the previous lessons, I think I performed (better / worse / equally well).

List two areas in this chapter in which you feel you could improve your skill or knowledge level:

1. _____
_____
_____
_____

2. _____
_____
_____
_____

Consider the following self-evaluation questions as they pertain to this chapter:

Am I taking effective notes?

Am I dedicating enough quality time to my studies?

Is anything distracting my focus?

Was any part of this chapter too advanced for me?

Did I find that I don't have enough background experience to sufficiently grasp certain subject areas?

For areas in which I did particularly well, was it because I'm particularly interested in that subject matter? How so?

Were some things easier to learn because I have prior experience in learning or working with the concepts or principles?

Did I find that certain portions of the textbook seem to be better organized and effective in explaining key points?

Selecting Personal Protective Clothing and Equipment

Chapter Orientation

Open the textbook to the *Selecting Personal Protective Clothing and Equipment* chapter. When you have finished looking through the chapter, respond to the following items. Use the textbook as you jot down your comments in the spaces provided below.

1. There were no NFPA standards for chemical-protective clothing at the time of the Benicia, California, chemical suit incident in 1983. Surprised? Have you ever considered what it would be like today if standards such as NFPA 1991, *Vapor Protective Ensembles for Hazardous Materials Emergencies*, had not been written?

   _____

   _____

   _____

   _____

2. Reflect on your current level of knowledge or background experience about the topics covered in this chapter. Where have you read about, learned about, or applied this knowledge in the past?

   _____

   _____

   _____

   _____

3. Which sections or parts of this chapter strike you as looking especially interesting?

   _____

   _____

   _____

   _____

   _____

4. Which particular subjects in this chapter are important for a person in your position to master?

   _____

   _____

   _____

   _____

   _____

5. What do you predict will be the most difficult things for you to learn in this chapter?

   _____

   _____

   _____

   _____

   _____

Learning Objectives

Examine the *Selecting Personal Protective Clothing and Equipment* chapter objectives and respond to the following question:

1. Which objectives in this chapter do you feel you can achieve right now, with a reasonable level of confidence?

_____

_____

_____

_____

As you read the sections of the chapter that deal with the objectives you have identified above, make sure your ideas and knowledge base match those of the authors. If they do not match, you should examine how your current understanding of the material differs from that of the authors. Depending on the level at which you wish to master the subject, discrepancies will have to be rectified and gaps will need to be filled.

Abbreviations and Acronyms

The following abbreviations and acronyms are used in the *Selecting Personal Protective Clothing and Equipment* chapter:

APR	Air-Purifying Respirator
ARFF	Aircraft Rescue Firefighting
ASTM	American Society of Testing and Materials
CBRN	Chemical, Biological, Radiological, and Nuclear
CPC	Chemical-Protective Clothing
CPE	Chlorinated Polyethylene
ESLI	End-of-Service-Life Indicator
MDPR	Minimum Detectable Permeation Rate
NIOSH	National Institute for Occupational Safety and Health
PAPR	Powered Air-Purifying Respirator
PNA	Polynuclear Aromatic Compounds
PPE	Personal Protective Clothing and Equipment
PVA	Polyvinyl Alcohol
PVC	Polyvinyl Chloride
RIT	Rapid Intervention Team
SAR	Supplied Air Respirator
SCBA	Self-Contained Breathing Apparatus
SDL	System Detection Limit
SEI	Safety Equipment Institute
SFC	Structural Firefighting Clothing
UL	Underwriters Laboratories

Abbreviations and Acronyms Crossword

Use the clues below to identify the correct abbreviation or acronym (from the preceding list) to solve the crossword.

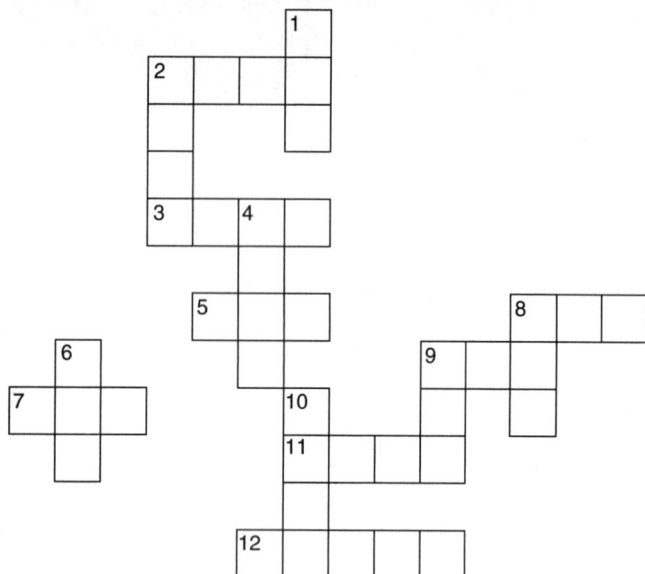

Across

2. Proximity suits are often needed for conducting these operations.

3. The minimum permeation rate that can be detected by the laboratory analytical system being used for the permeation test.

5. Filtration devices that remove particulate matter, gases, or vapors from the atmosphere.

7. Single or multi-piece garments that are specially designed and configured to protect the wearer's torso, head, arms, legs, hands, and other body parts.

8. The minimum amount of chemical breakthrough that can be detected by the laboratory analytical system being used for the permeation test.

9. Gloves made of this material provide excellent barrier protection against certain petroleum solvents but break down on exposure to water.

11. A component of the structural firefighting clothing ensemble.

12. A Federal agency that tests and certifies respiratory protective devices.

Down

1. Protective clothing normally worn by fire fighters during structural firefighting operations.

2. An organization that develops voluntary consensus standards for testing protective clothing.

4. Air-purifying respirators that use a blower to force the ambient air through air-purifying elements to a full-face mask.

6. The last line of defense that comes into play if your selection of tactical objectives and site safety procedures can't keep the bad stuff off you.

8. A positive-pressure respirator that is supplied by either an air-line hose or breathing air cylinders connected to the respirator by a short air line.

9. A product of combustion that has led to increased concerns about the contamination and decontamination of structural firefighting clothing

10. A device that warns the user of the approach of the end of adequate respiratory protection.

Study Session Overview

Use the textbook to help you answer the questions or work from memory.

1. What is the operational goal of "Select personal protective clothing and equipment?" (Hint: return to *The Eight Step Process©: An Overview* chapter if you need help.)

   _____

   _____

   _____

   _____

2. In your own words, explain what is meant by the statement that "PPE is NOT your first line of defense; it is your last line of defense."

   _____

   _____

   _____

   _____

3. Wearing any type and level of impermeable protective clothing creates the potential for heat stress injuries. Review the section on heat stress in the *Health and Safety* chapter. What are the signs and symptoms of heat cramps, heat exhaustion, and heat stroke?

   _____

   _____

   _____

   _____

4. The selection and use of specialized protective clothing at a hazmat emergency should be approached from a systems perspective. Briefly describe this system.

   _____

   _____

   _____

   _____

5. Assume you are the Hazmat Group Safety Officer or other Hazmat Group personnel (e.g., Entry Leader). You are about to conduct a pre-entry safety briefing. Which topics will you include in your briefing?

   _____

   _____

   _____

   _____

6. Which kinds of information should be recorded for all chemical-vapor-protective clothing?

   _____

   _____

   _____

   _____

Self-Test

Use the textbook to help you answer the questions or work from memory.

Select the best term from the following list to answer questions #1–7. You may write the word or its letter (a, b, or c) in the blank.

   a. Degradation

   b. Penetration

   c. Permeation

1. _____ is the physical destruction or decomposition of a clothing material due to exposure to chemicals, use, or ambient conditions.

2. _____ is the process by which a hazardous chemical moves through a given material on the molecular level.

3. _____ is the flow or movement of a hazardous chemical through closures, seams, porous materials, and pinholes or other imperfections in the material.

4. _____ can lead to protective clothing failures when breakthrough times are exceeded.

5. _____ can be caused by clothing material degradation, manufacturing defects, physical damage to the suit (e.g., punctures, abrasions), normal wear and tear, and PPE defects.

6. _____ resistance data are provided as "pass" or "fail" relative to the specific chemical or mixture tested.

7. _____ is noted by visible signs such as charring, shrinking, swelling, color change, or dissolving, or by testing the clothing material for weight changes, stiffening, loss of fabric tensile strength, and so on.

8. The following statements about breakthrough time are all true except one. Which statement is *false*?

   a. Breakthrough time is measured in units of minutes or hours.

   b. Breakthrough times are determined by laboratory testing procedures.

   c. Breakthrough time is defined as the time from the initial chemical attack on the outside of the material to its detection inside.

   d. Breakthrough time is the only reliable measure of chemical resistance.

9. Chemical permeation through protective clothing is a three-step process—absorption, diffusion, and:

   a. adsorption.

   b. desorption.

   c. disintegration.

   d. evaporation.

10. For reference purposes, a permeation rate of 0.9 $\mu/cm^2/min$ is equal to:

   a. 1 drop per minute.

   b. 1 dram per minute.

   c. 1 drop per hour.

   d. 1 milliliter per 12-hour period.

11. True / False: Once a chemical has begun the diffusion process, it may continue to diffuse even after the chemical itself has been removed from the outside surface of the material.

12. True / False: Chemical-protective clothing is not appropriate for firefighting operations or for protection in flammable or explosive environments.

13. The *most* critical factor in evaluating and choosing chemical protective clothing is:

    a. shelf life.

    b. chemical resistance.

    c. flexibility.

    d. decontamination and disposal.

14. The following statements about NFPA 1994, *Protective Ensemble for Chemical/Biological Terrorism Incidents*, are all true except one. Which statement is *false*?

    a. NFPA 1994 was originally enacted in 2001 as a result of the growing terrorism problem.

    b. Many of the NFPA1994 testing requirements are similar to those found in both NFPA 1991 and NFPA 1992.

    c. All NFPA 1994 ensembles (i.e., garment, gloves, and footwear) are reusable.

    d. NFPA 1994 defines three classes of ensembles based on the perceived threats at an incident.

15. The _____ Respirator Certification Requirements outline the requirements for particulate respirators.

    a. NIOSH

    b. ASTM

    c. EPA

    d. NFPA

Place an X next to each statement to indicate whether the statement pertains only to APRs (air purification respirators), PAPRs (powered-air purification respirators), or both.

APRs	PAPRs	
		16. Can be found with either a full-face or half-face configuration
		17. Are negative pressure respirators
		18. Have a protection factor of 50:1
		19. Have a protection factor of 1000:1
		20. Are rated according to filter efficiency degradation
		21. Cannot be used in IDLH environments
		22. Should not be used in the presence or potential presence of unidentified contaminants
		23. Are well-suited for operations involving solids, dusts, and powders

24. The following are all components of a supplied air respirator (SAR) *except*:

    a. air-line hose.

    b. source of breathing air.

    c. positive-pressure respirator.

    d. alkaline scrubber.

25. Structural firefighting clothing (SFC) is normally not the first PPE choice for most hazmat response scenarios, except when the incident involves:

    a. corrosives and PCBs.

    b. cyanide compounds.

    c. infectious bloodborne diseases.

    d. flammable gas and liquid fire incidents.

26. What is the minimum level of respiratory protection when wearing structural firefighting clothing in hazmat environments?

   a. Half-face air purification respirators

   b. Full-face air purification respirators

   c. Powered-air purification respirators

   d. Positive-pressure SCBA

27. When may liquid chemical splash-protective clothing be used?

   a. ___yes / ___no: When the vapors or gases present are not suspected of containing high concentrations of chemicals that are harmful to, or can be absorbed by, the skin.

   b. ___ yes / ___no: When it is highly unlikely that the user will be exposed to high concentrations of vapors, gases, or liquid chemicals that will affect any exposed skin areas.

   c. ___ yes / ___no: When operations will not be conducted in a flammable atmosphere.

28. When should chemical-vapor-protective clothing be used?

   a. ___ yes / ___no: When extremely hazardous substances are known or suspected to be present, and skin contact is possible.

   b. ___ yes / ___no: When anticipated operations involve unknown or unidentified substances and the scenario dictates that vapor-tight skin protection is required.

29. High-temperature protective clothing is designed primarily for _____ heat exposures.

   a. ambient

   b. conductive

   c. radiant

   d. None of the above

30. The two types of high-temperature protective clothing are: 1) proximity suits and 2):

   a. fire entry suits.

   b. flash overgarments.

   c. aluminized suits.

   d. insulation suits.

Practice

1. Develop procedures to address one or more of the following scenarios involving chemical vapor suits:

   - Loss of air supply
   - Loss of suit integrity
   - Loss of communications
   - Buddy down in the hot zone

Study Group Activity

1. In this exercise you will examine the advantages and disadvantages of limited-use garments. Half of the group should brainstorm the advantages, while the other half of the group brainstorms the disadvantages. Rejoin and compare notes.

2. The selection and use of specialized protective clothing at a hazmat emergency should include an evaluation of the capabilities of the user/wearer. Half of the group should develop a list of physical stressors that may affect responders; the other half of the group should develop a list of the psychological stressors. Rejoin, present the lists, and discuss ways to minimize the identified stressors.

Study Group Learning Through Inquiry Scenario 8–1

You are the emergency response supervisor within a refinery. Among the various processes within the refinery is a hydrofluoric acid alkylation (HF alky) unit. An emergency at the HF alky unit could potentially involve a wide range of mixtures of toxic hydrofluoric acid (HF) or flammable liquefied petroleum gas (LPG). Depending on the nature of the incident and the location of a release, you could be faced with a situation ranging from 100% LPG, to a 50% LPG/50% HF mixture, to 100% HF.

As part of developing a preincident plan for the unit, respond to the following personal protective clothing issues.

1. Which types of chemical-protective clothing would be compatible with the hydrofluoric acid?

2. What would your protective clothing recommendations be for the following scenarios?

   a. Release of 100% HF acid as a result of a line break during a transfer operation from a tank truck to the HF storage tank.

   b. Fire and release from a 2-inch pipe flange of a mixture containing approximately 75% LPG and 25% HF.

   c. Fire from an LPG line on the unit.

3. What are the advantages and limitations of combining chemical-protective clothing and thermal-protective clothing into a single ensemble?

4. Assume that your ERT responded to an incident where 75% LPG and 25% HF were involved. A member from your ERT was burned when the leak ignited. He was wearing a chemically compatible Level A fully encapsulating suit at the time of injury. In your opinion, was the ERT member wearing the correct type of PPE for the hazards present? If you answered no, do you think that OSHA would be justified in issuing a citation for failure to wear the proper protective clothing? As an Incident Commander working for an oil company, do you think that you could be held liable in a civil suit brought against you by the burned ERT member?

Study Group Learning Through Inquiry Scenario 8–2

You are a fire officer assigned to an engine company in a municipal fire department. You are responding to a report of an odor in the vicinity of the Acme Ice Plant. While responding to the call, your dispatcher informs you that the department has just received a 911 call for an ammonia leak inside the Ice Plant. A man is reported trapped. The assignment is upgraded to include three engine companies, a ladder company, a rescue squad, a BLS ambulance, and the HMRT. You are a member of the first-due company and about three minutes ahead of the second-due company.

You arrive on the scene and position your apparatus upwind of the incident. Approximately 10 employees are standing outside in the parking lot. Some have been exposed to the ammonia and are having difficulty breathing. Two people are vomiting.

A supervisor from the Ice Plant frantically explains that a maintenance crew was working on a liquid refrigeration valve when something went wrong. One of the maintenance crew members did not get out and is trapped inside the building.

The Battalion Chief arrives on-scene and orders your company to don SCBA and enter the building to search for the missing worker. You have received training to the OSHA First Responder Operations Level. Although you are not especially familiar with the hazards of ammonia, you know from a recent training class with the HMRT that it is toxic by inhalation, a skin irritant, and potentially explosive when released in confined areas. Your PPE meets the requirements of NFPA 1500, and consists of fire-retardant structural firefighting clothing with full-length pants, coat, hood, helmet, boots, gloves, and a PASS device.

Based on this situation and the information provided in this chapter, answer the following questions:

1. What are the hazards of ammonia? In your opinion, what is the level of risk based on the situation as it has been presented to you?

2. Are you wearing the proper type and level of protective clothing and equipment to enter the building and conduct search and rescue operations? If you answered yes, explain why. If you answered no, what would be the proper type of protective clothing and equipment?

3. How do you feel about the Battalion Chief's order to enter the building for search and rescue? Is this a reasonable order? If you believe that this is an unreasonable order, would you obey and enter the building even if you believed that your crew was being placed at an unreasonable risk?

4. How would you approach the Battalion Chief if you felt the risk was unreasonable? What would the consequences be if you refused to enter the building and the trapped maintenance worker died? What if you entered the building and a member of your crew died? What if no one died, the rescue was successful, and your crew was awarded a Unit Citation for Heroism—would this change your opinion about taking the risk? If so, why?

Summary and Review

1. Which of the following terms is defined as the physical destruction or decomposition of a clothing material due to exposure to chemicals, use, or ambient conditions?
   a. Disintegration
   b. Degradation

   c. Diffusion

   d. Degeneration

2. List at least two visible signs of degradation:

3. In which form are penetration resistance data provided?

   a. "Pass" or "fail"

   b. >480 minutes or >8 hours

   c. $\mu\gamma/cm^2/min$

   d. Grade A, B, or C

4. What is the definition of breakthrough time?

   a. The time it takes a hazardous chemical to physically destroy or decompose a chemical-protective clothing material

   b. The rate at which the chemical passes through a chemical-protective clothing material

   c. The time from the initial chemical attack on the outside of the material until its desorption and detection inside

   d. The rate of flow or movement of a hazardous chemical through closures, seams, porous materials, and pinholes or other imperfections in a material

5. How can you tell if an article of protective clothing is NFPA-compliant?

6. Why isn't degradation or immersion testing considered sufficient for compatibility assessment?

7. List at least two advantages of using SCBA:

8. List at least two disadvantages or limitations of using SCBA:

9. In an emergency response, liquid chemical splash-protective clothing is often used for initial response operations. Identify at least one other operation in which liquid chemical splash-protective clothing is used:

10. Identify one advantage of selecting a chemical vapor suit with a supplied-air respirator:

11. True / False: Proximity suits are not designed to offer any substantial chemical protection.

12. Procedures for donning and doffing of specific CPC ensembles should be based on:
    a. NFPA 1991 or NFPA 1992.
    b. the manufacturer's instructions.
    c. NIOSH Pocket Guide recommendations.
    d. the environmental conditions (e.g., weather, noise).

13. At a minimum, protective clothing should be inspected at certain benchmarks, including upon receipt from the manufacture or vendor. Identify at least two other benchmarks.

Self-Evaluation

Review all your work in this lesson and note your stronger and weaker areas.

Overall I feel I did (very well / well / fair / not so well) on the acronyms and abbreviations exercise.

Overall I feel I did (very well / well / fair / not so well) on the self-test questions.

Overall I feel I did (very well / well / fair / not so well) on the summary and review questions.

When compared to the previous lessons, I think I performed (better / worse / equally well).

List two areas in this chapter in which you feel you could improve your skill or knowledge level:

1. _____

   _____

   _____

   _____

2. _____

   _____

   _____

   _____

Consider the following self-evaluation questions as they pertain to this chapter:

Am I taking effective notes?

Am I dedicating enough quality time to my studies?

Is anything distracting my focus?

Was any part of this chapter too advanced for me?

Did I find that I don't have enough background experience to sufficiently grasp certain subject areas?

For areas in which I did particularly well, was it because I'm particularly interested in that subject matter? How so?

Were some things easier to learn because I have prior experience in learning or working with the concepts or principles?

Did I find that certain portions of the textbook seem to be better organized and effective in explaining key points?

Information Management and Resource Coordination

Chapter Orientation

Open the textbook to the *Information Management and Resource Coordination* chapter. When you have finished looking through the chapter, respond to the following items. Use the textbook as you jot down your comments in the spaces provided below.

1. What would you say is the key to coordinating information and resources required to resolve a working hazmat emergency?

   _____

   _____

   _____

   _____

   _____

2. Reflect on your current level of knowledge or background experience about the topics covered in this chapter. Where have you read about, learned about, or applied this knowledge in the past?

   _____

   _____

   _____

   _____

   _____

3. Which sections or parts of this chapter strike you as looking especially interesting?

   _____

   _____

   _____

   _____

4. Which particular subjects in this chapter are important for a person in your position to master?

   _____

   _____

   _____

   _____

5. What do you predict will be the most difficult things for you to learn in this chapter?

   _____

   _____

   _____

   _____

Learning Objectives

Examine the *Information Management and Resource Coordination* chapter objectives and respond to the following question:

1. Which objectives in this chapter do you feel you can achieve right now, with a reasonable level of confidence?

_____

_____

_____

_____

As you read the sections of the chapter that deal with the objectives you have identified above, make sure your ideas and knowledge base match those of the authors. If they do not match, you should examine how your current understanding of the material differs from that of the authors. Depending on the level at which you wish to master the subject, discrepancies will have to be rectified and gaps will need to be filled.

Abbreviations and Acronyms

The following abbreviations and acronyms are used in the *Information Management and Resource Coordination* chapter:

CHEMTREC®	Chemical Transportation Emergency Center
COP	Common Operating Picture
EOC	Emergency Operations Center
FRA	Federal Railroad Administration
HMRT	Hazardous Materials Response Team
IAP	Incident Action Plan
IC	Incident Commander
ICP	Incident Command Post
JIC	Joint Information Center
LEPC	Local Emergency Planning Committee
MOU	Memorandum of Understanding
NGO	Nongovernmental Organizations
NTSB	National Transportation Safety Board
OPSEC	Operations Security
PIO	Public Information Officer
PPE	Personal Protective Clothing and Equipment
WISER	Wireless Information System for Emergency Responders

Word Puzzle

This puzzle has two steps.

Step 1: Use the word clues below to identify each acronym or abbreviation from the preceding list. Write the answers in the boxes.

The individual responsible for establishing and managing the overall incident action plan.

An organized group of employees who are expected to perform work to handle and control actual or potential leaks or spills of hazardous substances requiring possible close approach to the substance.

The strategic goals, tactical objectives, and support requirements for the incident.

Secured site where government or facility officials exercise centralized direction and control in an emergency.

Equipment provided to shield or isolate a person from the chemical, physical, and thermal hazards that may be encountered at a hazardous materials incident.

The location at which the primary command functions are executed, usually co-located with the incident base.

Federal agency within the DOT concerned with intermodal transportation safety.

Written agreement between different organizations that may have overlapping areas of responsibility.

Serves as a focal point in the community for information and discussions about hazardous substances, emergency planning, and health and environmental risks.

Step 2: Take the letters you have written in the circles (only) above and unscramble them to reveal the term below.

							N	S		S				R		T	Y

Study Session Overview

Use the textbook to help you answer the questions or work from memory.

1. In your own words and using examples, explain how accurate, effective, and timely information management and resource coordination is essential to the following considerations:

    a. Safety of responders

    _____

    _____

    _____

    _____

b. Development of the IC's incident action plan

_____   _____

_____

_____

_____

_____

2. In your own words and using examples, explain how poor or ineffective information management and resource coordination can politically damage the IC's credibility and ultimately undermine the response operation.

_____

_____

_____

_____

_____

_____

3. According to this chapter, most resource coordination problems fall into three categories:

- Failure to understand or work within the incident command structure
- Given the type and nature of the incident, failure to anticipate potential problems and "gaps" in information or resources
- Communications and personality problems between the players

Based on your own experience, which of these occur most frequently? Which has the potential to create the greatest problems? Give some examples.

_____

_____

_____

_____

_____

_____

Self-Test

Use the textbook to help you answer the questions or work from memory.

1. Specific gravity, flash point, exposures values, and vapor density are all examples of:
   a. data.
   b. facts.
   c. information.
   d. opinion.

2. Statements made or observations about something that has occurred and has been verified and validated as being true are:
   a. data.
   b. facts.
   c. information.
   d. opinion.

3. True / False: Information management must begin well before the incident.

4. Which of the following is a common mistake made at the emergency scene regarding data?

   a. Looking up the wrong chemical in the database

   b. Not copying the information down correctly

   c. Failure to validate the data using another reference source

   d. All of the above

5. Which of the following organizations provides a standard on the type of information that should be included in preincident plans?

   a. National Fire Protection Association (NFPA)

   b. National Institute of Standards and Technology (NIST)

   c. Emergency Management Institute (EMI)

   d. Office of Force Readiness Deployment (OFRD)

6. Within NIMS, external agencies may fall into three categories: 1) assisting agencies; 2) cooperating agencies; and 3):

   a. liaison agencies.

   b. independent federal organizations.

   c. federally funded organizations.

   d. nongovernmental organizations.

7. When a Hazmat Group is organized within the ICS, the responsibility for hazmat information is usually delegated to the:

   a. Hazmat Group Supervisor.

   b. Logistics Section Chief.

   c. Liaison Officer.

   d. Situation Unit Leader.

8. To be effective, checklists must be:

   a. NFPA 472 compliant.

   b. updated on a regular basis.

   c. stored electronically.

   d. standardized across all neighboring jurisdictions.

9. Supply resources differ from equipment resources in that supply resources:

   a. are less expensive.

   b. are easier to requisition.

   c. are expendable.

   d. don't usually require decontamination.

10. Within the ICS organization, resources are coordinated and tracked by the:

   a. Staging Area Manager.

   b. Information Officer.

   c. Safety Officer.

   d. Resource Unit Leader.

11. Which of the following is *not* a unit of the Service Branch of the Logistics Section?

a. Communications

b. Facilities

c. Medical (ERP Rehab)

d. Food

Practice

1. What makes you suspect that the data provided below on chlorine are *not* correct? What should you do to verify these data?

<table>
<tr><td align="center">**CHLORINE**</td><td>**Hazard Rating = High**</td></tr>
<tr><td rowspan="3"></td><td>Fire Hazard Index = 3</td></tr>
<tr><td>Health Hazard Index = 3</td></tr>
<tr><td>Reactivity Hazard Index = 1</td></tr>
<tr><td align="center">CAS NO. 7774-50-5

UN/NA NO. 1017</td><td></td></tr>
</table>

PRODUCT NAME: Chlorine

SPECIAL HAZARDS: Chlorine is an extremely poisonous gas and is toxic by inhalation. It is a strong oxidizer and may react explosively or form explosive compounds with many common chemicals, such as hydrocarbons.

PHYSICAL AND CHEMICAL CHARACTERISTICS:

Exposure values	TLV/TWA = 0.5 ppm; OSHA PEL = 0.5 ppm; STEL = 1 ppm; IDLH = 30 ppm
Flash point	Not applicable
Flammable limits (% by volume)	Not applicable
Autoignition temperature	Not applicable
Boiling point	–29°F
Specific gravity	1.4 @ 68°F
Vapor density	2.5 @ –30°F

_____

_____

_____

_____

2. You have been assigned to write a proposal to obtain a set of published emergency response references, including reference manuals and guidebooks, online databases and information web sites, and a listing of technical information centers. Conduct the necessary background research and develop your proposal. Note: Your proposal should include an evaluation of each reference in terms of its value, usefulness, applicability to hazards and risks in your jurisdiction, user-friendliness, durability, accessibility, and other similar factors. Be sure to specify special equipment or training that will be needed to access or use the references. Your proposal should be comprehensive, realistic, and justifiable.

_____

_____

_____

_____

_____

3. For purposes of this scenario, assume that you are the Incident Commander.

On a Thursday morning in May at 11:00 A.M. your fire department responds to a report of a fire at a metal processing plant. Which information do you need prior to and upon arrival at the scene?

_____

_____

_____

_____

_____

The foreman who reported the fire tells you he smelled rubber burning, but when he went to investigate he could not see through the smoke. Based on this initial report, do you have data, facts, or opinion?

_____

_____

_____

_____

_____

Size-up confirms a large fire in the processing area. The preplan indicates that sulfuric acid is your primary concern. What are your information needs at this point? Which resources will you use to get the information you require?

_____

_____

_____

_____

_____

4. Based on your own experience, identify some of the advantages you have found in using a checklist system for coordinating and managing information in the field. Compare your list with the advantages identified in the *Information Management and Resource Coordination* chapter.

_____

_____

_____

_____

5. Which skills and abilities are needed to perform in the role of the hazmat information leader? Under which conditions would you feel qualified to function in this role?

_____

_____

_____

_____

6. For purposes of this scenario, assume that you are the Incident Commander at an airport.

At 8 A.M. you receive a dispatch to the International Terminal where more than 100 people are reported to be experiencing dizziness, nausea, vomiting, and respiratory distress. No fires have been detected, and an unconfirmed report said that there was no noticeable odors or smoke. Based on this limited information, which human, equipment, and supply resources do you anticipate needing for the first hour of this response?

_____

_____

_____

_____

_____

7. Which of the four resource coordination problems discussed in this chapter occur most frequently in your experience? Which is the most difficult to address? Support your position with examples.

_____

_____

_____

_____

_____

Study Group Activity

1. One member in the group should relate a personal experience at a hazmat incident in which he or she was initially overwhelmed by the quantity of incoming information. As a group, discuss the scenario and suggest ways in which the problem with information overload could have been better managed.

_____

_____

_____

_____

_____

2. Using the preplan criteria listed in this chapter and presented below, group members should identify problem areas in their own jurisdiction that may require special preplanning:

- Type of hazards and risks present
- Critical infrastructure
- Economically sensitive sites
- Environmentally sensitive exposures
- Unusual or poor water supply requirements
- Locations that will require large quantities of foam concentrate

- Restricted or delayed response routes
- Poor accessibility

_____

_____

_____

_____

_____

3. As a group, design a template that can be used for preplanning key transportation areas in your jurisdiction.

_____

_____

_____

_____

_____

4. As a group, identify potential problems that could be encountered with electronic information storage and retrieval under a variety of conditions. Then brainstorm solutions and work-arounds.

_____

_____

_____

_____

5. Have a member of the group recount an incident in which the use of empathy (as discussed in this chapter) helped build support and "buy-in" from a special-interest group or representative.

_____

_____

_____

_____

Study Group Learning Through Inquiry Scenario 9–1

Your Chief has asked you to develop recommendations for information management improvement for the new hazardous materials response unit that is being designed. The chief is a "gadget" kind of guy, and he wants the new unit to be state-of-the art with the latest communications and computer equipment. He wants "everything paperless" on the new unit.

1. Which criteria will you use to develop your information management specifications?

2. Based on your research, you have determined that the Chief's idea of being completely automated and "paperless" is not the best course of action. How would you make your case that there should be some redundancy in your information management system? Which references would you carry as hard copy versus soft copy?

3. What are the advantages of using checklists as a method of coordinating information at the incident scene?

Study Group Learning Through Inquiry Scenario 9–2

You are the on-duty acting military shift supervisor at a U.S. Air Force base fire department. You are a Technical Sergeant (E-6) with eight years' experience. It is a holiday, and there is very limited activity on the base. At 5:45 P.M. you receive an emergency call for a fuel spill inside of a hangar on the flight line. You respond with three aircraft rescue/firefighting (ARFF) vehicles and a structural engine company.

When you arrive on the scene, you find an F-16 fighter parked inside the hangar with about 10 gallons of jet fuel pooled on the hangar floor. The door is open on the hangar, and several maintenance vehicles are parked inside. You take a combustible gas indicator reading near the hangar door and determine that the atmosphere at floor level is approximately 80% of the LEL. You have ordered that the area be isolated and ignition sources secured; however, your prefire plan does not indicate where the overhead heating system master switch is located. You are very concerned about a fire and explosion.

There is no fixed foam system inside the hangar. You make the decision to foam the fuel spill down as a way to reduce the vapors and reduce the potential for an explosion inside the hangar. You are well aware of both the aircraft replacement costs and maintenance costs associated with your actions.

As you begin foaming down the fuel spill under the aircraft, the Maintenance Supervisor shows up. He outranks you as a First Lieutenant and expresses his opinion that it is a waste of time and money to foam down the spill. He doesn't want to make a big mess inside the hangar, and he wants to use fire department and maintenance personnel to manually tow the aircraft outside of the hangar. You feel that this strategy is neither safe nor practical.

Even though the First Lieutenant outranks you, you know that Air Force regulations place the emergency on-scene commander in charge of fire and rescue operations regardless of rank. In other words, you would be within your authority to foam the spill down regardless of the Maintenance Supervisor's objections. However, for obvious chain-of-command and career reasons, you don't want this situation to turn into a major political incident. This is also your first time as acting shift officer, and you don't want to blow any potential career opportunities.

Based on the background information provided and information in this chapter, answer the following questions.

1. Which types of resources would you need to safely mitigate this incident? Be specific and list both the equipment and human resources required.

2. In your opinion, what is the fundamental problem with this situation from an information and resource coordination perspective?

3. The Maintenance Supervisor is presenting a major political problem for you. This problem has the potential to become a major issue within the military chain of command if you handle it incorrectly. How would you resolve his concerns? Refer to the *Information Management and Resource Coordination* chapter and review the model solutions provided for resolving conflict. Which techniques would be most effective in helping you avoid a major conflict with the officer?

Summary and Review

1. Fill in the blank:

   Data + _____ = Information

2. Information management must begin well *before* the incident. For example, which type of information will be needed at the scene? How should the information be complied? Identify at least one other information management question that should be answered *before* an incident occurs:

3. One way to deal with information overload is to distinguish between information we need to know versus information that is nice to know. For example, we need to know what the hazards of the materials are and what the PPE requirements are. Identify at least two other "need-to-know" pieces of information:

4. Several information sources should be immediately accessible from the incident scene, such as preincident tactical plans. Identify at least one other source of information:

5. Which kinds of information can be found in a facility emergency response plan?

6. Which kinds of facilities will most certainly present accessibility problems to responding units and must be preplanned?

7. Fill in the boxes on the left side of the following organizational chart:

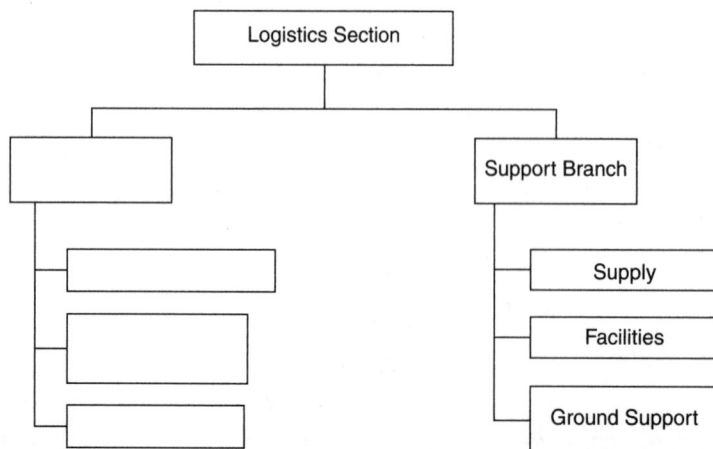

```
                    ┌─────────────────────┐
                    │  Logistics Section  │
                    └─────────────────────┘
              ┌──────────────┴───────────────────────┐
    ┌───────────────────┐              ┌─────────────────────┐
    │                   │              │   Support Branch    │
    └───────────────────┘              └─────────────────────┘
         │                                      │
    ┌────┴──────────────┐              ┌────────┴────────┐
    │                   │              │     Supply      │
    └───────────────────┘              └─────────────────┘
         │                                      │
    ┌────┴──────────────┐              ┌────────┴────────┐
    │                   │              │   Facilities    │
    └───────────────────┘              └─────────────────┘
         │                                      │
    ┌────┴──────────────┐              ┌────────┴────────┐
    │                   │              │  Ground Support │
    └───────────────────┘              └─────────────────┘
```

8. Most emergency response organizations call on the same cast of external players on a regular basis. What is the best way to ensure that these external resources will be properly coordinated and integrated into your command structure?

Self-Evaluation

Review all your work in this lesson and note your stronger and weaker areas.

Overall I feel I did (very well / well / fair / not so well) on the acronyms and abbreviations exercise.

Overall I feel I did (very well / well / fair / not so well) on the self-test questions.

Overall I feel I did (very well / well / fair / not so well) on the summary and review questions.

When compared to the previous lessons, I think I performed (better / worse / equally well).

List two areas in this chapter in which you feel you could improve your skill or knowledge level:

1. _____

_____

_____

_____

_____

_____

2. _____

_____

_____

_____

_____

_____

Consider the following self-evaluation questions as they pertain to this chapter:

Am I taking effective notes?

Am I dedicating enough quality time to my studies?

Is anything distracting my focus?

Was any part of this chapter too advanced for me?

Did I find that I don't have enough background experience to sufficiently grasp certain subject areas?

For areas in which I did particularly well, was it because I'm particularly interested in that subject matter? How so?

Were some things easier to learn because I have prior experience in learning or working with the concepts or principles?

Did I find that certain portions of the textbook seem to be better organized and effective in explaining key points?

Implementing Response Objectives

Chapter Orientation

Open the textbook to the *Implementing Response Objectives* chapter. When you have finished looking through the chapter, respond to the following items. Use the textbook as you jot down your comments in the spaces provided below.

1. In one or two sentences, what is this chapter about?

_____

_____

_____

_____

_____

2. Reflect on your current level of knowledge or background experience about the topics covered in this chapter. Where have you read about, learned about, or applied this knowledge in the past?

_____

_____

_____

_____

_____

_____

3. Which sections or parts of this chapter strike you as looking especially interesting?

_____

_____

_____

_____

_____

4. Which particular subjects in this chapter are important for a person in your position to master?

_____

_____

_____

_____

_____

5. What do you predict will be the most difficult things for you to learn in this chapter?

_____

_____

_____

_____

Learning Objectives

Examine the *Implementing Response Objectives* chapter objectives and respond to the following question:

1. Which objectives in this chapter do you feel you can achieve right now, with a reasonable level of confidence?

_____

_____

_____

_____

_____

As you read the sections of the chapter that deal with the objectives you have identified above, make sure your ideas and knowledge base match those of the authors. If they do not match, you should examine how your current understanding of the material differs from that of the authors. Depending on the level at which you wish to master the subject, discrepancies will have to be rectified and gaps will need to be filled.

Abbreviations and Acronyms

The following abbreviations and acronyms are used in the *Implementing Response Objectives* chapter:

AFFF	Aqueous Film-Forming Foam
API	American Petroleum Institute
ARC	Alcohol-Resistant (Foam) Concentrate
ARFF	Aircraft Rescue Firefighting
ASME	American Society of Mechanical Engineers
BLEVE	Boiling Liquid Expanding Vapor Explosion
CAP	Civil Air Patrol
CCTV	Closed-Circuit Television
CNG	Compressed Natural Gas
EERC	Ethanol Emergency Response Coalition
EFV	Excess Flow Valve
EOD	Explosive Ordnance Disposal
ESD	Emergency Shutdown
FFFP	Film-Forming Fluoroprotein Foam
LNG	Liquefied Natural Gas
LPG	Liquefied Petroleum Gas
MODU	Mobile Offshore Drilling Unit
MOTEL	Magnitude, Occurrence, Timing, Effects, and Location
NTSB	National Transportation Safety Board
PRV	Pressure-Relief Valve
QA/QC	Quality Assurance/Quality Control
SONS	Spill of National Significance

Abbreviations and Acronyms Exercise

Fill in the blank using the abbreviations and acronyms from the preceding list.

1. Required on bulk LPG storage tanks' liquid internal safety valve: _____

2. When applied to a polar solvent fuel, will often create a polymeric membrane rather than a film over the fuel: _____

3. Federal group responsible for investigating major transportation disasters: _____

4. Personnel who must be notified whenever an explosive device is suspected to be involved in an incident: _____

5. Often accompanied by a large fireball if a flammable gas is involved: _____

6. Sets industry and manufacturing codes and standards that enhance public safety: _____

7. Firefighting related to airport operations and aircraft safety: _____

8. Synthetic Class B firefighting foam consisting of fluorochemical and hydrocarbon surfactants: _____

9. Class B firefighting foam based on fluoroprotein foam: _____

10. Acronym for the five factors that should be evaluated by the IC to influence the outcome of the emergency: _____

11. If spilled on the ground, it will boil rapidly at first and then boil slowly as the ground cools: _____

Study Session Overview

Use the textbook to help you answer the questions or work from memory.

1. There has been a significant reduction in the number of hazmat incidents resulting in multiple fatalities and casualties over the last two decades. According to this chapter, what is one of the primary reasons for this reduction?

_____

_____

_____

_____

_____

2. In your own words, explain the concept of the emergency timeline, and describe its value in determining which strategy and tactics are best suited to change the outcome for a hazmat incident.

_____

_____

_____

_____

_____

3. Although offensive operations can increase the risk to emergency responders, when may the risk be justified?

_____

_____

_____

_____

_____

4. Some tactics can be used to delay events or slow down the clock until entry teams are ready to implement the solution to the problem. Give an example of a tactical option that could be used to buy time and explain how it would work:

_____

_____

_____

_____

_____

5. What makes technical rescue problems involving hazardous materials especially difficult to plan for and execute?

_____

_____

_____

_____

_____

6. What are some of the advantages of conducting confinement operations rather than containment operations?

_____

_____

_____

_____

7. What are the role and activities of public safety responders during product transfer and removal operations?

_____

_____

_____

_____

Self-Test

Use the textbook to help you answer the questions or work from memory.

1. The following statements all pertain to *strategy* except one. What is the exception?

   a. A strategy is a plan for managing resources.

   b. A strategy is usually very broad in nature.

   c. A strategy is selected at the Command Level.

   d. Only one strategy should be pursued at a time during an incident.

2. The following statements all pertain to *tactics* except one. What is the exception?

   a. Tactics are specific objectives used to achieve strategic goals.

   b. Tactics are normally decided at the section or group/division levels in the command structure.

   c. Tactics are rarely or never implemented when operating in the nonintervention mode.

   d. Several tactics may be implemented simultaneously during an incident.

3. Nonintervention means taking no action other than:

   a. rescue.

   b. isolating the area.

   c. fire control.

   d. incident stabilization.

4. What are the specific activities that accomplish a tactical objective called?

   a. Operations

   b. Tasks

   c. Procedures

   d. Priorities

5. With regard to the "First Law of Hot Zone Operations," assume you are "trained to play, dressed to play, and using the buddy system; you have a backup capability and an emergency decon capability, but you don't have command approval for the entry rescue operation." Can you work in the hot zone?

   Yes / No

6. Technical rescue includes rescue of one or more victims who have been exposed to the hazmat and require:

   a. triage.

   b. immediate relocation.

   c. physical extrication.

   d. treatment.

7. The following are all examples of confined spaces, as defined by OSHA *except*:

   a. aircraft cockpit.

   b. ventilation duct.

   c. boiler.

   d. pipeline.

Match each confinement tactical option in Column A to its description or definition in Column B.

Column A	Column B
Confinement Tactic	**Description/Definition**

_____   **8.** Absorption

A. Constructing a barrier across a waterway to stop/control the product flow and pick up the liquid or solid contaminants.

_____   **9.** Adsorption

B. Chemical method by which a water-soluble solution, usually a corrosive, is diluted by adding large volumes of water to the spill.

_____   **10.** Covering

C. Barriers are constructed on ground or placed in a waterway to intentionally control the movement of a hazardous material into an area where it will pose less harm to the community and the environment.

_____   **11.** Damming

D. Fans or water spray is used to disperse or move vapors away from certain areas or materials.

_____   **12.** Diking

E. Physical process of absorbing or "picking up" a liquid hazmat to prevent enlargement of the contaminated area.

_____   **13.** Dilution

F. Physical method of confinement to reduce or eliminate the vapors emanating from a spilled or released material.

_____   **14.** Diversion

G. Placing a tarp over a spill of dust or powder.

_____   **15.** Dispersion

H. The chemical process in which a sorbate (liquid hazardous material) interacts with a solid sorbent surface.

_____   **16.** Retention

I. A liquid is temporarily contained in an area where it can be absorbed, neutralized, or picked up for proper disposal.

_____   **17.** Vapor dispersion

J. Chemical and biological agents are used to disperse or break up the material involved in liquid spills on water.

_____   **18.** Vapor suppression

K. Barriers are constructed on ground used to control the movement of liquids, sludges, solids, or other materials.

19. The following are all <u>physical</u> methods of confinement *except*:

a. covering.

b. damming.

c. dispersion.

d. vapor suppression.

20. Which of the following confinement tactics should be used only when all other reasonable methods of mitigation and removal have proved unacceptable?

a. Dilution

b. Diversion

c. Dispersion

d. Covering

21. All of the following statements about leak control strategies and containment tactics are all true except one. What is the exception?

a. Containment tactics are implemented only when defensive options would be too expensive or time consuming.

b. Containment tactics should be approved only after conducting a thorough hazard and risk evaluation.

c. Containment tactics require personnel to enter the hot zone to control the release at its source.

d. Containment tactics should be considered high-risk operations.

22. For alkali spills, what is the most widely favored neutralizing agent from an environmental perspective?

    a. Ascorbic acid

    b. Acetic acid

    c. Hydrochloric acid

    d. Perchloric acid

23. True / False: Overpack containers must be labeled in accordance with DOT hazmat regulations if they will be transported from the scene.

24. When used as a method of containment, plugs must be compatible with both the chemical and the:

    a. available tools.

    b. available time.

    c. container.

    d. atmospheric conditions.

25. Which of the following is a critical factor in evaluating the application and use of patching tactics?

    a. Container size

    b. Container pressure

    c. Container position

    d. Availability of properly rated commercial patches

26. Responders should consult with container specialists to assess the level of risk and control options for which of the following types of liquid cargo tank truck leaks?

    a. Valve leaks

    b. MC-306/DOT-406 piping leaks

    c. Liquid being released from pressure-relief devices

    d. Breaches in the vapor space

27. Which of the following is a drawback of flaring to reduce or control pressure?

    a. Flares are designed to burn only vapor product.

    b. Flaring can cause dangerous pressure buildups in other locations.

    c. Flaring often takes a very long time to accomplish.

    d. Flaring can weaken the structural integrity of the container.

28. Which of the following is often used to contain small releases of liquid mercury?

    a. Vacuuming

    b. Solidification

    c. Neutralization

    d. Absorption

29. Which of the following tactical options for flammable liquid emergencies is sometimes implemented when there is an insufficient water supply?

    a. Nonintervention

    b. Offensive tactics

    c. Defensive tactics

    d. Defensive tactics directed only at primary exposures

30. With regard to tank firefighting, AFFF foam:

    a. is available in 12% concentrations.

    b. is compatible with Purple K dry chemical agent.

    c. should not be used with salt water.

    d. is not suitable for subsurface injection.

31. Which of the following standards recommends minimum foam application rates for specific fuels, foams, and applications?

    a. NFPA 326

    b. NFPA 30

    c. NFPA 11

    d. NFPA 58

32. With regard to cooling water requirements, pressure vessels should have a minimum of _____ gpm applied at the point of fire impingement.

    a. 250

    b. 300

    c. 400

    d. 500

33. Pressure-fed flammable gas fires may produce direct flame impingement on nearby vessels and cause catastrophic tank failure within _____ of exposure.

    a. 5 to 20 minutes

    b. 20 to 30 minutes

    c. 30 to 60 minutes

    d. 1 to 5 hours

34. BLEVE is an acronym for boiling liquid _____ vapor explosion.

    a. evaporating

    b. escaping

    c. expanding

    d. ejecting

35. True / False: Never extinguish a pressure-fed flammable gas fire unless you can control the fuel supply.

36. Liquefied natural gas (LNG) is _____ in its liquid state.

    a. propane

    b. butane

    c. ethane

    d. methane

37. What is the expansion ratio for LNG?

    a. 100:1

    b. 250:1

    c. 600:1

    d. 1,200:1

38. All of the following statements about the hazard and risk process for reactive chemicals are true except one. What is the exception?

  a. Small quantities of highly reactive chemicals can pose significant risks.

  b. Incidents involving reactive chemicals will typically require the expertise of technical information and product specialists who are familiar with the materials involved.

  c. The type of container will vary depending on whether the chemical is a raw material, an intermediate material being used to form another chemical or product, or the finished product.

  d. Unlike flammable gases containers, reactive chemical containers do not have pressure-relief devices.

39. Product removal operations cannot commence until after the incident site is stabilized. What does stabilization mean?

  a. All fires have been extinguished.

  b. Ignition sources have been controlled.

  c. All spills and leaks have been controlled.

  d. All of the above

40. True / False: Bonding and grounding must be established before product removal and transfer operations can begin.

41. Gravity flow as a method of liquid product transfer is often ineffective for:

  a. liquids containing alcohol.

  b. viscous liquids.

  c. flammable liquids.

  d. corrosive liquids.

42. Gas transfers are based on the basic principle that materials will naturally flow from:

  a. high-pressure to low-pressure areas.

  b. high elevation to low elevation.

  c. high-concentration to low-concentration areas.

  d. volume to a vacuum.

Practice

1. Review the fire emergency timeline presented in the chapter. Using this timeline as a model, create and label an example of a hazardous materials emergency timeline.

_____

_____

_____

_____

2. Briefly explain and give a real-world example of how each of the following defensive tactical options can help buy time until the most effective offensive tactic can be implemented.

  a. Barriers:

_____

_____

_____

_____

b. Distance (i.e., separating people from the hazmat):

_____

_____

_____

_____

_____

c. Time (i.e., reducing the duration of the release):

_____

_____

_____

_____

_____

d. Techniques (i.e., procedures to stop the leak):

_____

_____

_____

_____

_____

3. Develop a detailed equipment and supplies inventory for handling a wide range of hazmat incidents using confinement tactics.

_____

_____

_____

_____

_____

4. Approximately 30 gallons of diesel fuel has been spilled into a fresh water lake at a marina. Which containment tactic would you use? Explain why.

_____

_____

_____

_____

5. Assume you are responding to an acid spill of approximately 1 gallon. The spill has a pH of 4. You want to dilute the spill to a pH of 6 for safer disposal. How many gallons of water will you need to do this?

_____

_____

_____

_____

6. How much soda ash (sodium carbonate) is needed to neutralize a 1-gallon spill of 70% sulfuric acid (specific gravity 1.8)?

7. A gasoline spill covers an area of 100 feet × 100 feet. You will be using Class B, 3% × 3% AFFF. You have 500 gallons of foam concentrate. Is this enough?

_____

_____

_____

_____

_____

8. Briefly describe a scenario in which you would conduct an aggressive leak control strategy versus a spill control strategy. Explain the factors in your scenario that would justify an offensive operation.

_____

_____

_____

_____

_____

9. The *Implementing Response Objectives* chapter presents a number of significant hazmat emergencies. If you were asked to make a conference presentation on one of the incidents, which one would you choose and why? Use other resources and research that incident in more detail.

_____

_____

_____

_____

_____

Study Group Activity

1. One person in the group should present a scenario from his or her own experience (or a case study) in which one of the following factors added to the uncertainty of the decision-making process:

   • Conflicting or uncertain information
   • Conflicting or competing values

   As a group, discuss the outcome and what can be learned from the scenario.

2. Review the discussion of the Texas City, TX, 1947 ammonium nitrate explosion in this chapter. Which parallels exist between this disaster and the threat that terrorism may present to our present and future critical infrastructure?

Study Group Learning Through Inquiry Scenario 10–1

You respond to a flammable liquid storage tank fire at a petroleum marketing terminal. The 150-ft internal floating roof tank was struck by lightning after being filled from a pipeline. When you arrive on the scene, you observe the following:

- The 150-ft tank has a full surface fire, and there are several small pool fires within the diked area.
- The fixed roof has been blown off into an adjoining 120-ft-diameter cone roof tank. Although the adjoining tank of #2 fuel oil is not yet burning, the side of the 40-ft-high tank has been punctured approximately 10 ft from the top, and product is flowing into the dike.
- Prefire plans have identified that water supplies are adequate in this area.
- There is approximately 2,500 gallons of AFFF 3%/3% firefighting foam concentrate available within 1 hour in the region. There are also numerous 1½-inch and 2½-inch foam handline nozzles available, although there are no foam cannons. An ARFF crash truck is available from a nearby Air National Guard base.

Based on the background information provided and the information in this chapter, answer the following questions.

1. What would be your strategic goals and tactical objectives for this fire? Would you use offensive tactics and attempt to attack and extinguish the fire, or would you use defensive tactics? Explain and justify your reason for using these tactics.

2. Do you have sufficient foam concentrate on hand to attack and extinguish the fire? If you believe that you do, how did you arrive at this conclusion? For example, if the facility operations manager does not believe that there is not enough foam available to extinguish the fire, how would you back up your opinion that there is?

3. What would be the short- and long-term environmental, public affairs, and political risks if the decision were made to pump out the burning tank and let the residual product burn itself out?

Study Group Learning Through Inquiry Scenario 10–2

You are the Safety Officer on the scene of an overturned tractor trailer incident on a secondary road. The truck contains a mixed load of hazardous materials in various small packages. The containers include the following:

- Ten 50-lb bags of fertilizer grade ammonium nitrate
- Ten 20-gallon containers of 35% concentration hydrogen peroxide
- Twenty 50-lb bags of low-grade potassium permanganate
- Fourteen 20-gallon containers of muriatic acid
- Fifteen 55-gallon drums of industrial-grade floor cleaner
- Seventeen 30-lb fiberboard containers of calcium hypochlorite

During the incident, your HMRT entered the hazard area and overpacked two leaking containers of muriatic acid and repackaged several bags of ammonium nitrate. You are now standing by for a carrier to arrive at the scene with a hazardous materials contractor to complete off-loading the vehicle and clean-up of the accident site.

Approximately 4 hours into the incident, a carrier representative arrives on the scene. Rather than hire a recognized environmental clean-up contractor, he has brought six dock workers who will be used to off-load the vehicle into another trailer.

Based on the information that has been provided and the information in this chapter, answer the following questions.

1. As the safety officer, what are your priorities at this point? Be specific and explain why the priorities you have selected are important to the overall safety of the operation.

2. Based on your priorities, are there any concerns that you have as far as clean-up is concerned? How would you express these concerns to your Incident Commander?

3. Which types of questions would you want to ask the carrier representative and the dock workers?

4. If you declare the incident as being terminated and leave the scene, what liability would there be to you as the Safety Officer if the dock workers were injured? Which steps could you recommend to reduce your liability?

Summary and Review

1. Five factors should be evaluated by the IC to positively influence the outcome of an emergency. These can be remembered by the acronym MOTEL. Identify the missing factors:

   M   Magnitude

   O   _____

   T   Timing

   E   Effects

   L   _____

2. The terms "strategy" and "tactics" are sometimes used interchangeably, but they actually have very different meanings. What is a strategy?

   a. A decision to take action

   b. A plan for managing resources

   c. A plan for visualizing outcomes

   d. A set of actions for achieving objectives

3. Common strategic goals implemented at hazmat incidents include rescue and public protective actions. Identify at least one other strategic goal:

4. Which of the following persons should *not* get involved in making detailed tactical-level decisions, such as which type of plug to use?

   a. Incident Commander (IC)

   b. Hazmat Group Leader

   c. Group/Division Leader

   d. Sector Officer

5. Selecting the best strategic goal involves weighing what will be gained against the "costs" of what will be lost—a process often easier said than done. Determining what will be "gained" involves weighing many different variables, including potential casualties and fatalities. Identify at least one other variable:

6. High-angle rescue is one example of a technical hazmat rescue situation. Identify another technical hazmat rescue situation:

7. Confined spaces have a number of hazardous characteristics, including limited egress. Identify at least two other hazardous characteristics:

8. Four important criteria must be met before dilution is attempted. One is determining in advance that the substance is not water reactive. Identify at least two of the other criteria:

9. Why are containment tactics considered offensive operations?

10. The following are all pressure isolation and reduction tactics *except*:
    a. flaring.
    b. vent and burn.
    c. vapor dispersion.
    d. scrubbing.

11. A number of size-up issues must be addressed at flammable liquids emergencies, including the time the fire started. Which other information will you want to have?

12. Under which conditions are offensive tactics implemented at flammable liquids fires?

13. True / False: Finished foams of a similar type but from different manufacturers (e.g., all AFFFs) are considered compatible.

14. With regard to cooling water for exposed tanks and pressure vessels, atmospheric storage tanks from 100-ft diameter to 150-ft diameter require _____ gallons of water per minute.

15. With regard to flammable gas emergencies, thermal stress is the primary concern in a fire situation; however, _____ stress may be an equal concern.

16. Tactical priorities for managing a flammable gas fire are to protect primary and secondary exposures to the fire and isolate the flammable gas source feeding the fire. Identify at least one additional tactical priority:

17. When a BLEVE is imminent, what is the best tactical strategy?

18. BLEVEs of bulk containers and process vessels can produce severe fire and fragmentation risks within _____ feet of the failed container.

   a. 500

   b. 1,000

   c. 1,500

   d. 3,000

19. Identify the two primary hazards of LPG:

20. When dealing with large quantities of strong oxidizers and organic peroxides, responders should consider treating the incident like a(n) _____ fire.

   a. flammable gas

   b. flammable liquids

   c. explosives

   d. radioactive materials

21. Before transferring site control to nonemergency response personnel, the IC should verify that any environmental spill contractors used for clean-up and recovery operations are trained per the requirements of:

   a. ANSI Z88.2, *Practices for Respiratory Protection.*

   b. NFPA 471, *Recommended Practice for Responding to Hazardous Material Incidents.*

   c. OSHA 1910.120, *Hazardous Waste Operations and Emergency Response.*

   d. OSHA Directive Number CPL 2-2/59A, *Inspection Procedures for the Hazardous Waste Operations and Emergency Response Standard.*

22. Identify at least one site safety consideration that should be addressed during product removal operation:

23. Of the three primary methods of liquid product transfer—gravity flow, pump transfer, and pressure transfer—which does *not* allow the use of vapor recovery?

24. Liquefied gases (MC-331 cargo tank trucks) and cryogenic liquids (MC-338 cargo tank trucks) may be transferred through the use of pumps, compressors, and pressure differential. Which of these methods does *not* increase the internal pressure of the damaged tank?

25. Depending on their rating and design features, vacuum trucks can handle:

   a. flammable and combustible liquids.

   b. corrosives.

   c. some poisons.

   d. All of the above

26. True / False: *Never* upright a loaded aluminum-shell MC-306/DOT-406 cargo tank truck.

27. The decision to either (1) offload the contents and then upright the container, or (2) upright the container while still loaded will depend on a number of variables, including the type of cargo tank truck involved. Identify at least one other variable that should be considered:

28. List at least three tactics for protecting evidence at a crime scene in a hazardous environment:

Self-Evaluation

Review all your work in this lesson and note your stronger and weaker areas.

Overall I feel I did (very well / well / fair / not so well) on the acronyms and abbreviations exercise.

Overall I feel I did (very well / well / fair / not so well) on the self-test questions.

Overall I feel I did (very well / well / fair / not so well) on the summary and review questions.

When compared to the previous lessons, I think I performed (better / worse / equally well).

List two areas in this chapter in which you feel you could improve your skill or knowledge level:

1. _____

_____

_____

_____

2. _____

_____

_____

_____

Consider the following self-evaluation questions as they pertain to this chapter:

Am I taking effective notes?

Am I dedicating enough quality time to my studies?

Is anything distracting my focus?

Was any part of this chapter too advanced for me?

Did I find that I don't have enough background experience to sufficiently grasp certain subject areas?

For areas in which I did particularly well, was it because I'm particularly interested in that subject matter? How so?

Were some things easier to learn because I have prior experience in learning or working with the concepts or principles?

Did I find that certain portions of the textbook seem to be better organized and effective in explaining key points?

Decontamination

Chapter Orientation

Open the textbook to the *Decontamination* chapter. When you have finished looking through the chapter, respond to the following items. Use the textbook as you jot down your comments in the spaces provided below.

1. In your own words, what is the basis for the statement, "If you are good at decon, you will probably be good at preventing exposures."

_____

_____

_____

_____

_____

2. Reflect on your current level of knowledge or background experience about the topics covered in this chapter. Where have you read about, learned about, or applied this knowledge in the past?

_____

_____

_____

_____

_____

3. Which sections or parts of this chapter strike you as looking especially interesting?

_____

_____

_____

_____

_____

4. Which particular subjects in this chapter are important for a person in your position to master?

_____

_____

_____

_____

_____

5. What do you predict will be the most difficult things for you to learn in this chapter?

_____

_____

_____

_____

_____

Learning Objectives

Examine the *Decontamination* chapter objectives and respond to the following question:

1. Which objectives in this chapter do you feel you can achieve right now, with a reasonable level of confidence?

_____

_____

_____

_____

_____

As you read the sections of the chapter that deal with the objectives you have identified above, make sure your ideas and knowledge base match those of the authors. If they do not match, you should examine how your current understanding of the material differs from that of the authors. Depending on the level at which you wish to master the subject, discrepancies will have to be rectified and gaps will need to be filled.

Abbreviations and Acronyms

The following abbreviations and acronyms are used in the *Decontamination* chapter:

ALARA	As Low as Reasonably Achievable
ART	Animal Response Team
Decon	Decontamination
HEPA	High-Efficiency Particulate Air
mR	Milliroentgens
PERO	Post-Emergency Response Operations
PID	Photo-ionization Detectors
RCRA	Resource Conservation and Recovery Act
SBCCOM	U.S. Army Soldier and Biological Chemical Command
START	Simple Triage and Rapid Treatment Process

Abbreviations and Acronyms Crossword

Use the clues below to identify the correct abbreviation or acronym (from the preceding list) to solve the crossword.

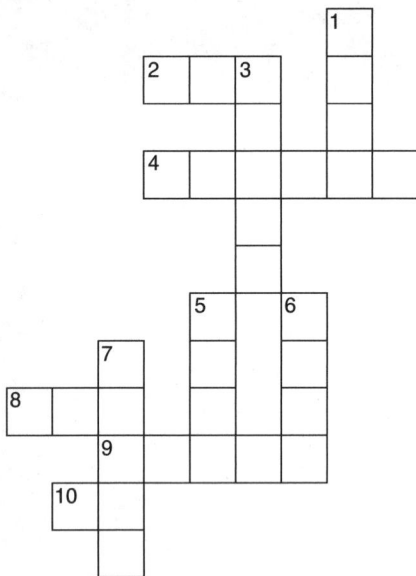

Across

2. Monitoring device used to determine the total concentration of many organic and some inorganic gases and vapors in the air.

4. Organization that produces *Guidelines for Mass Casualty Decontamination During a Terrorist Chemical Agent Incident*.

8. Group that addresses contaminated animal handling and response issues.

9. A principle intended to ensure that most radiation exposures will be well below the defined limit.

10. A unit of radiation exposure.

Down

1. That portion of an emergency response performed after the immediate threat of a release has been stabilized or eliminated and clean-up of the site has begun.

3. The removal of hazardous substances from people and equipment to the extent necessary to preclude foreseeable health effects.

5. Type of filter that can capture particles as small as 0.1 micron.

6. Gives the EPA the authority to control hazardous waste.

7. A triage protocol designed to be used for victims of acute traumatic injuries.

Study Session Overview

Use the textbook to help you answer the questions or work from memory.

1. At which point in the Eight Step Process© must you begin considering decon methods and procedures?

_____

_____

_____

_____

2. Give an example that illustrates how a responder who has been contaminated while wearing PPE could also be exposed to the contaminant.

_____

_____

_____

_____

3. Explain the difference between surface contaminants and permeation contaminants. Provide an example of each.

_____

_____

_____

_____

4. Why are product specialists often the best source of decon information for public safety organizations?

_____

_____

_____

_____

5. Why should a contaminated victim's clothing should be removed directionally, from head to toe?

_____

_____

_____

_____

6. Which considerations or concerns must be addressed during decon at a scene involving possible criminal activity?

_____

_____

_____

_____

Self-Test

Use the textbook to help you answer the questions or work from memory.

1. The scope, nature, and complexity of decon operations are defined by:
   a. the hazards and risks presented by the incident.
   b. the type and level of protective clothing worn.
   c. the decon equipment available.
   d. the extent to which responders have been exposed.

2. True / False: You cannot conduct safe entry operations if you have no way to perform decontamination.

3. What is the first rule of decontamination?

   a. Avoid contamination.

   b. Conduct all operations in the Hot Zone.

   c. Begin clean-up operations as soon as possible.

   d. Use disposable protective clothing.

4. Which of the following terms is defined as the initial phase of the decontamination process during which the amount of surface contaminant is significantly reduced?

   a. Mass decontamination

   b. Direct decontamination

   c. Gross decontamination

   d. Primary decontamination

5. Emergency decontamination is the process of immediately reducing contamination of individuals in potentially life-threatening situations with or without:

   a. standby medical assistance.

   b. standard decon procedures.

   c. the acknowledgment of the incident commander.

   d. the formal establishment of a decontamination corridor.

6. Which of the following is an example of an exposure?

   a. A responder breathes toluene vapors but does not experience any symptoms

   b. A bystander gets gasoline on his hands

   c. A child swallows a small amount of mercury

   d. All of the above

7. Use the following words to complete the sentence: exposure, decon, contaminant.

   If contact with the _____ can be controlled, the risk of _____ is reduced, and the need for _____ can be minimized.

8. True / False: Permeation can occur with any porous material, not just PPE.

9. Secondary contamination occurs when:

   a. the contaminant is not completely removed from a material and continues to permeate through the material.

   b. any form of exposure occurs as a result of a breach or failure of PPE.

   c. a person who is already contaminated makes contact with a person or object that is not contaminated.

   d. a person who has been contaminated causes another person to become exposed.

10. Mercury, ethylene dibromide (EDB), and benzene are all examples of:

    a. persistent contaminants.

    b. allergenic contaminants.

    c. reactive contaminants.

    d. chronic toxicity contaminants.

11. Which of the following is another name for embryotoxic contaminants?

    a. Teratogens

    b. Nerve agents

    c. Poisons

    d. Tertiary hazards

12. What is virulence?

    a. The speed at which viruses reproduce and spread

    b. The ability of the biological material to cause disease

    c. The degree to which viruses are resistant to antibodies

    d. The potential to cause damage to the human body as a result of a single exposure

13. The passage of _____ half-lives will bring a radiation level down to 1% of what it is at the time you take the first reading.

    a. 2

    b. 4

    c. 7

    d. 10

Identify each of the following decon methods as either a physical method or a chemical method.

14. Physical Method / Chemical Method: Dilution

15. Physical Method / Chemical Method: Brushing and scraping

16. Physical Method / Chemical Method: Pressurized air

17. Physical Method / Chemical Method: Disinfection

18. Physical Method / Chemical Method: Adsorption

19. Physical Method / Chemical Method: Sterilization

20. Physical Method / Chemical Method: Washing

21. Physical Method / Chemical Method: Neutralization

22. Physical Method / Chemical Method: Solidification

23. Physical Method / Chemical Method: Evaporation

24. The following statements about absorption are all true except one. What is the exception?

    a. Absorbent materials should be inert.

    b. Contaminants in absorbents remain chemically unchanged.

    c. Absorption has universal and nearly boundless application for decontaminating personnel.

    d. The most readily available absorbents are soil, diatomaceous earth, and vermiculite.

25. Who does the Decon Leader report to?

    a. Incident Commander

    b. Safety Officer

    c. Hazmat Group Supervisor

    d. Entry Officer

26. The following statements about the decontamination corridor are all true except one. What is the exception?
    a. The decon corridor should be clearly identified.
    b. Only one decon corridor should be operating at a time.
    c. The decon corridor begins in the hot zone and exits near the Warm Zone/Cold Zone interface.
    d. Portable tents and specially designed vehicles or trailers may be integrated into the decon corridor.

27. Which of the following persons should receive secondary decon?
    a. Decon personnel
    b. Entry team members
    c. Victims requiring follow-up medical treatment or evaluation
    d. Persons contaminated with flammable materials

28. During decon, which of the following should always be the last item removed?
    a. Gloves
    b. Boots
    c. Undergarments
    d. Respiratory protection

29. When dealing with large numbers of victims in a mass decon situation, which of the following is *crucial* and should be accomplished as soon as possible?
    a. Quick air monitoring and detection
    b. Establishment of the mass decon corridor
    c. Transportation of all victims to medical facilities
    d. Separation of significantly contaminated individuals from those who may not be contaminated

30. Studies involving chemical warfare agents show that victims should be washed for at least 3 minutes but no longer than:
    a. 5 minutes.
    b. 10 minutes.
    c. 15 minutes.
    d. 20 minutes.

Practice

1. The *Decontamination* chapter advises responders to always expect flammables and combustibles to present more than one contamination problem. Using outside resources, identify at least one secondary hazard and one tertiary hazard for each of the following: gasoline, acetone, benzene, and ethanol (liquids).

   _____

   _____

   _____

   _____

2. Arrange to visit a fixed facility in which hazardous materials will be found, such as a manufacturing facility. During your site visit, identify at least three fixed or engineered safety systems that may be found within such facilities and describe how each can facilitate the delivery of timely and effective decon.

_____

_____

_____

_____

3. Write an SOP or checklist for clean-up operations for decon of small and portable equipment, such as hand tools, fire hose, and monitoring equipment.

_____

_____

_____

_____

4. Write an SOP or checklist for clean-up/decon of motor vehicles and heavy equipment.

_____

_____

_____

_____

Important Terminology

The following are all important terms that you should know.

Safe refuge area
Contaminant
Contamination
Decontamination
Decontamination corridor
Emergency decontamination
Degradation
Disinfection
Exposure
Sterilization

On the line next to each term in Column A, print the letter of its definition from Column B.

Column A	**Column B**
Terminology	**Definitions**
_____ 1. Safe refuge area	**A.** The molecular breakdown of the spilled or released material to render it less hazardous.
_____ 2. Contaminant	**B.** The physical and/or chemical process of reducing and preventing the spread of contamination from persons and equipment used at a hazardous materials incident.

Column A	Column B
Terminology	**Definitions**

_____ **3.** Contamination

C. The process of destroying all microorganisms in or on an object.

_____ **4.** Decontamination

D. The physical process of immediately reducing contamination of individuals in potentially life-threatening situations with or without the formal establishment of a decontamination corridor.

_____ **5.** Decontamination corridor

E. A hazardous material that physically remains on or in people or equipment, thereby creating a continuing risk of direct injury or a risk of exposure outside of the hot zone.

_____ **6.** Emergency decontamination

F. The process used to destroy the majority of recognized pathogenic microorganisms.

_____ **7.** Degradation

G. A distinct area within the warm zone that functions as a bridge between the hot zone and the cold zone, where decontamination stations are located.

_____ **8.** Disinfection

H. A temporary holding area for contaminated people until a decontamination corridor is set up.

_____ **9.** Exposure

I. The process of transferring a hazardous material from its source to people or equipment, which may act as a carrier.

_____ **10.** Sterilization

J. The process by which people and equipment are subjected to or come in contact with a hazardous material.

Study Group Activity

1. Assume you are responding to an emergency that involves leaking drums of oleum. Set your own parameters for the incident, and then, working individually, each member of the group should list his or her decon method(s) and concerns. Then rejoin to discuss as a group. If time permits, repeat this activity by substituting drums of magnesium phosphide for the oleum.

2. As a group, discuss and elaborate on the following statement found in this chapter: "The fact that an area is contaminated with an etiological contaminant does not necessarily mean that a person has been exposed or is susceptible to the effects of an exposure."

3. Download and read the National Alliance of State Animal and Agricultural Emergency Program's (NASAAEP) *Animal Decontamination Best Practices* at http://www.learn.cfsph.iastate.edu/dr/wg.docs/decon-whitepaper6-23-12.pdf (or search the terms "NASAAEP Best Practices"). Then discuss the paper as a group, focusing on challenges and planning considerations pertinent to your own locality.

Study Group Learning Through Inquiry Scenario 11–1

An emergency medical technician (EMT) and paramedic are staffing an EMS unit, which has been dispatched to a single-family dwelling for a man suffering a heart attack. The individual is found in the basement slumped over a table. On the table and floor are several chemical containers, including those holding herbicides and sodium hypochlorite. The individual was known to "play with chemical mixtures" in his home and was apparently in the process of developing a chemical formulation when he was stricken.

While handling the patient, both EMS personnel become unknowingly contaminated by some of the liquid product lying on the table. As the EMS personnel arrive at the hospital, they begin to experience a burning sensation on their arms and legs.

You are the Hazmat Officer and have been asked to respond to the hospital to provide technical advice. Although several emergency room personnel have received some hazardous materials training, the hospital has not yet developed a procedure for handling chemically contaminated patients and protecting their own personnel.

Based on the information provided and the information in this chapter, answer the following questions.

1. Has the patient been exposed or contaminated? What is the difference? Have the paramedics been exposed or contaminated? How could you tell the difference? How about the hospital workers? Have they been exposed or contaminated?

2. Which options are available as to where the patient should be located in the hospital? What are the pros and cons of each option?

3. How would you handle and treat both the chemically contaminated patient and the EMS personnel? Which precautions would you take to protect yourself during treatment?

4. Unfortunately, the patient dies shortly after his arrival at the hospital. Due to the unknown circumstances surrounding the incident, an autopsy is ordered. As the Hazmat Officer, you are asked to provide health and safety recommendations for conducting the autopsy to ensure that the hospital staff are properly protected during the estimated 2-hour event. Which recommendations would you provide?

Study Group Learning Through Inquiry Scenario 11–2

You are the HMRT team leader at the scene of a hazmat incident involving a major poison gas release at a chemical process facility. A decision has been made to take immediate offensive action to close a valve that would stop the flow of gas. There is concern that the gas will drift downwind and go beyond the plant property and affect a nearby nursing home and medical facility.

A two-person entry team wearing chemical vapor protective clothing (EPA Level A) enters the contaminated area to close the valve. Both HMRT members and the backup crew have an internal suit radio communication capability.

The entry team locates the correct valve but needs a crows foot wrench to get the leverage they need to close the valve. Not having the proper wrench and running low on air, the Hazmat Safety Officer orders them out of the area to change out their air cylinders. As the entry team approaches the decontamination area, one member indicates by radio that he is experiencing a tingling situation in his fingers and toes. His voice is tense, and he is clearly excited and breathing heavily. His partner is in control and is not experiencing any difficulties.

Based on the information that you have been provided and the information in this chapter, answer the following questions.

1. What could be the problem with the entry team member who is experiencing the tingling sensation in his fingers and toes? Of the potential problems that you described, which ones are life threatening?

2. How would you handle the decontamination operation when the entry team arrives at the decon station? Does this situation justify an emergency decontamination? If you conducted an emergency decon, what risks would there be to the entry team? What risks would there be to the decon team?

3. Assuming that you had a second team standing by for immediate entry with the crows foot wrench, would you permit them to enter the hazard area and close the valve while you are decontaminating the first entry team? If so, why? If not, why would you delay entry of the second team?

Summary and Review

1. OSHA 1910.120 defines decontamination as the removal of hazardous substances from employees and their equipment to the extent necessary to:

   a. eliminate all potential exposure.

   b. preclude foreseeable health effects.

   c. reduce the risk of exposure outside of the hot zone.

   d. make the hazardous material inert.

2. Which of the following methods of decontamination is intended to immediately reduce contamination of a person in a potentially life-threatening situation with or without the formal establishment of a decontamination corridor?

   a. Technical decontamination

   b. Gross decontamination

   c. Physical decontamination

   d. Emergency decontamination

3. Which types of decon are normally conducted in support of emergency responder recon and entry operations at a hazardous materials incident, as well for contaminated victims requiring medical treatment?

   a. Mass decontamination

   b. Technical decontamination

   c. Radiological decontamination

   d. Cross-decontamination

4. Explain how each of the following factors can influence permeation:

• Contact time:

• Concentration of the contaminant:

• Temperature:

• Physical state of the contaminant:

5. Which of the following is a water-reactive contaminant?

  a. Magnesium phosphide

  b. Isocyanate

  c. Formamine

  d. Lewisite

6. Explain how "dose" influences an etiologic or biological material's ability to invade and alter the human body:

7. Which of the following is a limitation of dry decon?

  a. It must be preceded by brushing or scraping.

  b. It can cause aerosolization of the contaminant into the surrounding atmosphere.

  c. There is increased potential for secondary contamination.

  d. It is not effective on petroleum-based materials.

8. Which of the following is primarily used to decontaminate equipment, vehicles, and structures that are contaminated with a corrosive material?

  a. Neutralization

  b. Dilution

  c. Evaporation

  d. Absorption

9. Is sterilization essentially the same as disinfection? If not, what is the difference?

10. Using visual observation, what are some indicators that decon is being effective or ineffective?

11. The Decon Leader normally performs all of the following activities *except*:

  a. determining the appropriate level of decontamination to be provided.

  b. coordinating the transfer of decontaminated patients requiring medical treatment.

  c. monitoring the effectiveness of decon operations.

  d. decontaminating all decon team personnel.

12. Briefly describe an ideal outdoor decon site:

13. Identify at least one problem that can force the relocation of the decon area from its initial site:

14. The level of skin and respiratory protection required by Decon Team members will be dependent upon: 1) the type of contaminants involved; 2) the level of contamination encountered by entry personnel; and 3):

15. Some online databases provide decon information. Identify at least three other sources for determining the appropriate decon methods:

16. Studies by the U.S. Army Soldier and Biological Chemical Command (SBCCOM) of nerve agents using harmless simulants to track contaminants have shown that approximately 80% of contaminants can be removed by:
  a. undressing.
  b. washing with plain water.
  c. washing with water and detergent.
  d. wiping with isopropyl alcohol.

17. What roles does law enforcement play in assisting with mass decon?

18. Information on designing decon capability within a health care facility can be found in which standard?
  a. NFPA 58
  b. NFPA 70
  c. NFPA 99
  d. NFPA 473

19. What is the purpose of including water/sewage treatment facilities in the development of equipment decon plans?

20. Identify at least one special precaution that should be taken in the event that contaminated materials must remain at the incident scene until they can be removed for off-site cleaning or disposal:

Self Evaluation

Review all your work in this lesson and note your stronger and weaker areas.

Overall I feel I did (very well / well / fair / not so well) on the acronyms and abbreviations exercise.

Overall I feel I did (very well / well / fair / not so well) on the self-test questions.

Overall I feel did (very well / well / fair / not so well) on the terminology exercise.

Overall I feel I did (very well / well / fair / not so well) on the summary and review questions.

When compared to the previous lessons, I think I performed (better / worse / equally well).
List two areas in this chapter in which you feel you could improve your skill or knowledge level:

1. _____
   _____
   _____
   _____

2. _____
   _____
   _____
   _____

Consider the following self-evaluation questions as they pertain to this chapter:

Am I taking effective notes?

Am I dedicating enough quality time to my studies?

Is anything distracting my focus?

Was any part of this chapter too advanced for me?

Did I find that I don't have enough background experience to sufficiently grasp certain subject areas?

For areas in which I did particularly well, was it because I'm particularly interested in that subject matter? How so?

Were some things easier to learn because I have prior experience in learning or working with the concepts or principles?

Did I find that certain portions of the textbook seem to be better organized and effective in explaining key points?

Terminating the Incident

Chapter Orientation

Open the textbook to the *Terminating the Incident* chapter. When you have finished looking through the chapter, respond to the following items. Use the textbook as you jot down your comments in the spaces provided below.

1. In your own words, how does a properly terminated hazmat incident help to ensure scene safety?

   _____

   _____

   _____

   _____

   _____

2. Reflect on your current level of knowledge or background experience about the topics covered in this chapter. Where have you read about, learned about, or applied this knowledge in the past?

   _____

   _____

   _____

   _____

   _____

3. Which sections or parts of this chapter strike you as looking especially interesting?

   _____

   _____

   _____

   _____

   _____

4. Which particular subjects in this chapter are important for a person in your position to master?

   _____

   _____

   _____

   _____

5. What do you predict will be the most difficult things for you to learn in this chapter?

   _____

   _____

   _____

   _____

Learning Objectives

Examine the *Terminating the Incident* chapter objectives and respond to the following question:

1. Which objectives in this chapter do you feel you can achieve right now, with a reasonable level of confidence?

_____

_____

_____

_____

_____

As you read the sections of the chapter that deal with the objectives you have identified above, make sure your ideas and knowledge base match those of the authors. If they do not match, you should examine how your current understanding of the material differs from that of the authors. Depending on the level at which you wish to master the subject, discrepancies will have to be rectified and gaps will need to be filled.

Abbreviations and Acronyms

The following abbreviations and acronyms are used in the *Terminating the Incident* chapter:

AAR	After Action Review
IC	Incident Commander
ICP	Incident Command Post
IP	Improvement Plan
LEPC	Local Emergency Planning Committee
NFIRS	National Fire Incident Reporting System
NRC	National Response Center
OPSEC	Operations Security
PERO	Post-Emergency Response Operations
PIA	Postincident Analysis
RQ	Reportable Quantity
SOPs	Standard Operating Procedures

Abbreviations and Acronyms Exercise

Each of the abbreviations and acronyms in the preceding list are used in the sentences below. Some sentences are true statements; others are false. For each sentence, indicate whether it is true or false.

1. True / False: The requirements of the OSHA HAZWOPER regulation (29 CFR 1910.120) clearly delineate between the emergency phase of an incident response and the PERO.

2. True / False: The IC should ensure that response operations are fully coordinated with law enforcement or investigation agencies involved in the incident.

3. True / False: The PERO Incident Commander may be a contractor.

4. True / False: The Safety Officer is responsible for transferring command to the PERO Incident Commander.

5. True / False: The IC is nearly always the best facilitator for the debriefing.

6. True / False: The PIA assures that the incident has been properly documented.

7. True / False: The PIA should not be released to outside agencies.

8. True / False: Most major public fire departments participate in the NFIRS.

9. True / False: NFIRS is sponsored by the National Fire Protection Association.

10. True / False: Lessons learned that have been identified through the critique process should be incorporated into a formal Improvement Plan (IP) for the emergency response system.

11. True / False: When multiple agencies or jurisdictions were involved in mitigating the incident, the AAR development and writing process may require a multi-agency working group.

12. True / False: Liability problems can occur when written SOPs are not followed in the field.

Study Session Overview

Use the textbook to help you answer the questions or work from memory.

1. Provide several examples of how scene safety can deteriorate when an incident transitions from its emergency phase to the restoration and recovery phase.

   _____

   _____

   _____

   _____

   _____

2. Which key consideration determines whether you should still be operating in the emergency response mode versus when it is time to transition to the restoration and recovery phase?

   _____

   _____

   _____

   _____

   _____

3. Termination involves five distinct activities. What are they, and in which order should they take place?

_____

_____

_____

_____

_____

4. From memory, list at least five sources of factual information about a response that would be collected in developing the post-incident analysis. Compare your list to the suggested sources of information listed in the chapter.

_____

_____

_____

_____

_____

5. Would you make a good critique leader? Why or why not?

_____

_____

_____

_____

Self-Test

Answer the following questions. You may use the text to help you answer the questions, or work from memory.

1. The following statements about termination are all true _except_:
  a. Termination is the final step in the Eight Step Incident Management Process©.
  b. Termination is the transition between the conclusion of emergency phase operations and the initiation of decontamination.
  c. Terminating the incident includes conducting a debriefing, post-incident analysis, and the incident critique.
  d. Termination includes the transfer of on-scene command from the IC to the individual responsible for managing the post-emergency response operations.

2. Which of the following would be identified in an effective debriefing?
  a. The need for a critique
  b. The need for more advanced training
  c. Poor or inadequate cooperation among responders
  d. Damaged equipment requiring servicing

3. The need for a critical incident stress debriefing is accessed during the:

   a. debriefing.

   b. incident notification.

   c. post-incident analysis.

   d. critique.

4. Ideally, debriefings should take place:

   a. after the critique.

   b. after the incident has been properly documented and reported to the appropriate regulatory agencies.

   c. as soon as the emergency phase of the operation is completed.

   d. as soon as the post-emergency response operations are concluded.

5. Which of the following topics should be addressed *first* during the debriefing?

   a. Health information

   b. Potential legal issues

   c. Assignment of a follow-up contact person

   d. Assignment of a critique leader

6. The post-incident analysis is conducted to:

   a. meet National Fire Incident Reporting System requirements.

   b. gather necessary information for the debriefing.

   c. prevent documentation of potentially damaging information.

   d. assure that the incident has been properly documented and reported to the appropriate regulatory agencies.

7. The Post-Incident Analysis Coordinator:

   a. is usually appointed during the on-scene debriefing.

   b. has the authority to determine who will have access to information.

   c. cooperates with other official investigators to reconstruct the incident completely.

   d. All of the above

8. The rough draft PIA report should be reviewed by _____ to verify that the available facts are arranged properly and actually took place.

   a. the PIA Coordinator

   b. the Incident Commander

   c. key responders

   d. official investigators

9. Under CERCLA, the responsible party must report to the _____ any spill or release of a specified hazardous substance in an amount equal to or greater than the reportable quantity (RQ) specified by EPA.

   a. State Emergency Response Commission (SERC)

   b. National Fire Incident Reporting System (NFIRS)

   c. National Response Center (NRC)

   d. Occupational Safety and Health Administration (OSHA)

10. The primary purpose of a critique is to:

    a. comply with OSHA requirements.

    b. develop recommendations for improving the emergency response system.

    c. develop a chronological review of who did what, when, and where during the incident.

    d. provide a foundation for the development of formal investigations, which are usually conducted to establish the probable cause of the accident.

11. Which of the following statements about the critique process is true?

    a. The Safety Officer should always lead the critique.

    b. The critique process should never be used to find fault with the performance of individuals.

    c. Critiques that are longer than 15 minutes are probably too long.

    d. Critique reports are normally confidential and should not be released to the public.

Practice

1. As noted in the textbook, debriefings should be no longer than 15–20 minutes long—yet an effective debriefing needs to cover a number of points. If you were in charge of debriefing a major hazmat incident, what would you do to help keep on track?

    _____

    _____

    _____

    _____

    _____

2. Write a letter addressed to a potential critique leader outside the response organization that briefly explains the critique process and outlines the role and duties of a critique leader.

    _____

    _____

    _____

    _____

    _____

3. Design a simple form that can be used to guide the IC in a face-to-face transfer of command briefing with the PERO Incident Commander.

    _____

    _____

    _____

    _____

    _____

4. Browse the NFRS web pages on the USFA web site (http://www.usfa.fema.gov/fireservice/nfirs/index.shtm). Who are your state's Points of Contact (POC)?

_____

_____

_____

_____

_____

5. It is important that lessons learned which have been identified through the critique process be converted into changes and improvements to the emergency response system. In what ways can your organization track and make sure that action items have been addressed?

_____

_____

_____

_____

_____

Study Group Activity

1. Discuss ways in which a strong critique program can potentially outweigh possible liability vulnerability.

2. Assume that all of the students, as a group, will be participating in a critique in which a major confrontation among the players is anticipated. Discuss and decide on a strategy for diffusing the problem as much as possible before the actual critique session takes place.

Study Group Learning Through Inquiry Scenario 12–1

You are the Safety Officer at a confined space incident that involved two contractors who died while attempting to clean sludge out of the bottom of an ethylene dibromide (EDB) tank. The first worker entered the tank without respiratory protection and was immediately overcome. The second worker crawled through the manway without respiratory protection to rescue his partner. He was also overcome by the EDB. When rescue personnel arrived on the scene 5 minutes later, they determined that the workers were discovered missing at lunch. A co-worker was sent to investigate and found both men trapped in the tank.

When it became obvious that the workers had been trapped for several hours before the 911 call was placed, the Incident Commander stopped rescue operations and treated the incident as a body recovery operation. A confined space rescue team conducted the body recovery operation without any unusual problems. The Incident Commander has asked you to coordinate the termination phase of the incident and conduct an on-site debriefing of fire and rescue personnel. He reminds you that ethylene dibromide is a bad actor and a known carcinogen. He wants to make sure that the incident is terminated correctly with the proper documentation.

Based on the information you have been provided and the information in this chapter, answer the following questions:

1. Which post-incident safety and health concerns do you have that need to be worked into the debriefing?

2. How long do you think it will take to conduct a debriefing that will adequately cover all of the points that you will need to make? Where would you conduct the debriefing, and who should attend? Considering the potential health effects of EDB, what are your thoughts on the need for follow-up medical evaluation? Would you conduct a critical incident stress debriefing? Why or why not?

3. How would you handle a question from a member of the confined space rescue team if he stated that he knew that EDB was a strong poison but he did not realize during the pre-entry safety briefing that it was carcinogenic? He specifically wants to know whether the rescue team should go to the hospital for a follow-up medical examination.

Study Group Learning Through Inquiry Scenario 12–2

Your organization has traditionally conducted formal critiques of major incidents, although many of the personnel feel that the critiques have been inadequate and often "gloss over" critical problems and issues that have occurred. After attending a local Safety Symposium, which featured a workshop on critiquing techniques, you have the opportunity to brief your supervisor on your concerns with the current critique system. During the discussion, you express your opinion that the manner in which critiques are conducted and findings and recommendations are developed, managed, and tracked is not very effective. In short, you tell your manager that the current critique process stinks and that the overall emergency response system is not being improved as a result of the critiques.

After hearing your viewpoints, your supervisor replies, "If you think the system is so bad, *you* develop a proposal to fix it!" You now have the opportunity you have been waiting for. Using the background information provided and the information discussed in this chapter, answer the following questions.

1. What should the primary emphasis of your critique program be? Why?

2. Considering your dissatisfaction with the existing critique program, who would you recommend as the best type of critique facilitator? For example, should the critique facilitator come from the safety department, be internal to your organization, be external to the organization, or come from somewhere else?

3. How do you propose to document and distribute the lessons learned from the critiques throughout the organization?

4. How should follow-up items be tracked? How would you handle a situation where a particular supervisor doesn't believe in the critique system and simply throws away the critique reports and fails to share them with his subordinates?

Summary and Review

1. When the IC terminates the emergency response phase and formally transfers command to the PERO Incident Commander, the IC should conduct a transfer briefing that covers the nature of the emergency and actions taken to stabilize and resolve the emergency. Identify at least two other matters that should be covered in the briefing:

2. Many agencies and individuals have a legitimate need for information about significant hazmat incidents, such as shipping representatives. Identify at least two other groups that have a legitimate need for such information:

3. The post-incident analysis (PIA) is the reconstruction of the incident to establish a clear picture of the events that took place during the emergency. It is conducted to assure that the incident has been properly documented and reported to the right regulatory agencies. Identify at least one other reason why a PIA is conducted:

4. The PIA should focus on six key topics, listed below. For each topic, identify the key questions that should be answered by the PIA.

   Example:

   - **Command and control**: Was the Incident Management System established and was the emergency response organized according to the existing emergency response plan and/or SOPs? Did information pass from section personnel to the incident commander or through appropriate channels? Were response objectives clearly communicated to field personnel at the task level?

   - **Tactical operations**:

   - **Resources**:

   - **Support services**:

   - **Plans and procedures**:

   - **Training**:

5. Listed below are five primary reasons for liability problems in emergency response work. They are worth considering as a case for building a strong critique program. Identify issues associated with each.

   Example:

   - **Problems with planning**: Plans and procedures are poorly written, out-of-date, and unrealistic. In addition, what is written in the SOP is not followed in the field.
   - **Problems with training**:

   - **Problems with identification of hazards**:

- **Problems with duty to warn**:

- **Problems with negligent operations**:

6. The following statements pertain to either debriefings or critiques. Next to each statement, indicate whether it primarily pertains to a debriefing or a critique.

   a. Debriefing / Critique: The single most important way for an organization to self-improve over time.

   b. Debriefing / Critique: Assigns information gathering responsibilities for a post-incident analysis.

   c. Debriefing / Critique: OSHA requires that one be conducted for every hazardous materials emergency response.

   d. Debriefing / Critique: Should begin as soon as the emergency phase of the operation is completed.

   e. Debriefing / Critique: Inform responders exactly which hazardous materials they were (potentially) exposed to and signs and symptoms of such exposure.

   f. Debriefing / Critique: Identify unsafe site conditions that will impact the clean-up and recovery phase.

   g. Debriefing / Critique: The primary purpose is to develop recommendations for improving the emergency response system and responder safety.

   h. Debriefing / Critique: Share information among emergency response organizations.

   i. Debriefing / Critique: Should last no longer than 15–20 minutes.

   j. Debriefing / Critique: May have a published agenda.

   k. Debriefing / Critique: Is followed up with a short report.

   l. Debriefing / Critique: Also known as an After Action Review (AAR) or in the military as a "Hot Wash."

Self-Evaluation

Review all your work in this lesson and note your stronger and weaker areas.

Overall I feel I did (very well / well / fair / not so well) on the acronyms and abbreviations exercise.

Overall I feel I did (very well / well / fair / not so well) on the self-test questions.

Overall I feel I did (very well / well / fair / not so well) on the summary and review questions.

When compared to the previous lessons, I think I performed (better / worse / equally well).

List two areas in this chapter in which you feel you could improve your skill or knowledge level:

1. _____

_____

_____

_____

_____

2. _____

_____

_____

_____

_____

Consider the following self-evaluation questions as they pertain to this chapter:

Am I taking effective notes?

Am I dedicating enough quality time to my studies?

Is anything distracting my focus?

Was any part of this chapter too advanced for me?

Did I find that I don't have enough background experience to sufficiently grasp certain subject areas?

For areas in which I did particularly well, was it because I'm particularly interested in that subject matter? How so?

Were some things easier to learn because I have prior experience in learning or working with the concepts or principles?

Did I find that certain portions of the textbook seem to be better organized and effective in explaining key points?